MATERIALS ENGINEERING

AUTHOR

Dr. Dheerendra Kumar Dwivedi
Professor,
Dept. of Mechanical & Industrial Engineering
IIT Roorkee

REVIEWER

Dr. Santosh Kumar
Professor,
Dept. of Mechanical Engineering (ME)
IIT BHU

KHANNA BOOK PUBLISHING CO. (P) Ltd.
4C/4344, Ansari Road, Darya Ganj,
New Delhi, India

All India Council for Technical Education
Nelson Mandela Marg, Vasant Kunj,
New Delhi, India

Dear Readers,

To prevent the piracy, this book is secured with HIGH SECURITY HOLOGRAM on the front title cover. In case you don't find the hologram on the front cover title, please write us to at contact@khannabooks.com or whatsapp us at +91-99109 09320 and avail special gift voucher for yourself.

Specimen of Hologram on front Cover title:

Moreover, there is a SPECIAL DISCOUNT COUPON for you with EVERY HOLOGRAM.

How to avail this SPECIAL DISCOUNT:

Step 1: Scratch the hologram

Step 2: Under the scratch area, your "coupon code" is available

Step 3: Logon to www.khannabooks.com

Step 4: Use your "coupon code" in the shopping cart and get your copy at a special discount

Step 5: Enjoy your reading!

ISBN: 978-93-5538-988-6

Book Code: UG420EN

Materials Engineering

by Dr. Dheerendra Kumar Dwivedi

First Edition: 2026

Originally Published by:

All India Council for Technical Education, Ministry of Education, GOI

Exclusively Published & Distributed by:

Khanna Book Publishing Co. (P) Ltd.

4C/4344, Ansari Road, Daryaganj,
New Delhi (INDIA)

Visit us at: **www.khannabooks.com**

Write us at: **contact@khannabooks.com**

To view complete list of books, Please scan the QR Code: KPH

Printed in Bharat.

प्रो. म. जगदीश कुमार
अध्यक्ष
Prof. M. Jagadesh Kumar
Chairman

सत्यमेव जयते

आज़ादी का अमृत महोत्सव

अखिल भारतीय तकनीकी शिक्षा परिषद्
(भारत सरकार का एक सांविधिक निकाय)
(शिक्षा मंत्रालय, भारत सरकार)
नेल्सन मंडेला मार्ग, वसंत कुंज, नई दिल्ली-110070
दूरभाष : 011-26131498
ई-मेल : chairman@aicte-india.org

ALL INDIA COUNCIL FOR TECHNICAL EDUCATION
(A STATUTORY BODY OF THE GOVT. OF INDIA)
(Ministry of Education, Govt. of India)
Nelson Mandela Marg, Vasant Kunj, New Delhi-110070
Phone : 011-26131498
E-mail : chairman@aicte-india.org

FOREWORD

Engineers are the backbone of the modern society. It is through them that engineering marvels have happened and improved quality of life across the world. They have driven humanity towards greater heights in a more evolved and unprecedented manner.

The All India Council for Technical Education (AICTE), led from the front and assisted students, faculty & institutions in every possible manner towards the strengthening of the technical education in the country. AICTE is always working towards promoting quality Technical Education to make India a modern developed nation with the integration of modern knowledge & traditional knowledge for the welfare of mankind.

An array of initiatives have been taken by AICTE in last decade which have been accelerate now by the National Education Policy (NEP)2020. The implementation of NEP under the visionary leadership of Hon'ble Prime Minister of India envisages the provision for education in regional languages to all, thereby ensuring that every graduate becomes competent enough and is in a position to contribute towards the national growth and development through innovation & entrepreneurship.

One of the spheres where AICTE had been relentlessly working since 2021-22 is providing high quality books prepared and translated by eminent educators in various Indian languages to its engineering students at Under Graduate & Diploma level. For the second year students, AICTE has identified 87 books at Under Graduate and Diploma Level courses, for translation in 12 Indian languages - Hindi, Tamil, Gujarati, Odia, Bengali, Kannada, Urdu, Punjabi, Telugu, Marathi, Assamese & Malayalam. In addition to the English medium, the 1044 books in different Indian Languages are going to support to engineering students to learn in their mother tongue. Currently, there are 39 institutions in 11 states offering courses in Indian languages in 7 disciplines like Biomedical Engineering, Civil Engineering, Computer Science & Engineering, Electrical Engineering, Electronics & Communication Engineering, Information Technology Engineering & Mechanical Engineering, Architecture, and Interior Designing. This will become possible due to active involvement and support of universities/institutions in different states.

On behalf of AICTE, I express sincere gratitude to all distinguished authors, reviewers and translators from different IITs, NITs and other institutions for their admirable contribution in a very short span of time.

AICTE is confident that these out comes based books with their rich content will help technical students master the subjects with factor comprehension and greater ease.

(Prof. M. Jagadesh Kumar)

Acknowledgement

The authors are grateful to the authorities of AICTE, particularly Prof. M. Jagadesh Kumar, Chairman; Prof. M. P. Poonia, Vice-Chairman; Prof. Rajive Kumar, Member-Secretary and Dr Amit Kumar Srivastava, Director, Faculty Development Cell for their planning to publish the books on Materials Engineering. We sincerely acknowledge the valuable contributions of the reviewer of the book Prof. Santosh Kumar, Professor, IIT BHU. Author wishes to acknowledge the assistance of Mr. Chirag Panwariya, Mr. Pankaj, Kaushik, Mr. Niwas Kumar Roy, Mr. Saurabh and Mr. M V Rao in compiling technical content and figures of chapter 6.

This book is an outcome of various suggestions of AICTE members, experts and authors who shared their opinion and thought to further develop the engineering education in our country. Acknowledgements are due to the contributors and different workers in this field whose published books, review articles, papers, photographs, footnotes, references and other valuable information enriched us at the time of writing the book.

Dheerendra Kumar Dwivedi

Preface

The book titled "Materials Engineering" is an outcome of the teaching and & R & D experience in field of materials process and manufacturing. The main behind writing this book is to expose fundamental of materials science, testing and evaluation of mechanical properties, and develop understanding on structure-properties relation and ways to modify the mechanical properties by controlling the microstructure using approaches like alloying, work hardening, thermal treatment. Keeping in mind the purpose of wide coverage as well as to provide essential supplementary information, topics recommended by AICTE have been in a very systematic and orderly manner throughout the book. Efforts have been made to explain the fundamental concepts of the subject using suitable schematic diagrams as far as possible.

The content of the manuscript has been prepared considering international standard text-books, handbooks besides with research and development experience of the author. Additionally, each unit is supported with few sections like questions for self-assessment, unsolved subjective and numerical questions and activities for further information, references for further reading etc. Apart from illustrations and examples as required, the book has been enriched with numerous problems in every unit for proper understanding of the related topics.

The book entitled "Materials Engineering" is largely based on model curriculum of AICTE for second year under-graduate students, which is equally suitable for under-graduate programs on mechanical, production, manufacturing engineering of other University and Institutions for subjects related to materials engineering. The book comprises six units in following sequence a) fundamentals of materials science, mechanical properties of materials, theories of failures under static loading and fatigue behaviour, phase diagrams and metallurgical transformation, heat treatment of steels and common engineering metals. It is important to note that in all units, the relevant laboratory practical have been included.

The present book "Materials Engineering" is meant to provide a thorough grounding on presented topic related to materials engineering which will enable future graduate engineers understand the factors affecting materials properties and how to modify the same if required in light of engineering applications. The subject matters are presented in a constructive manner so that graduate engineers are better prepared to work in real world at the very forefront of technology.

I sincerely hope that the book will inspire the students to learn and discuss the ideas behind basic principles of materials engineering and will surely contribute to the development of a solid foundation of the subject. I would be thankful to all beneficial comments and suggestions which will contribute to the improvement of the future editions of the book. It gives us immense pleasure to place this book in the hands of the teachers and students. It was indeed a big pleasure to work on different aspects covering in the book.

Dheerendra Kumar Dwivedi

Outcome Based Education

For the implementation of an outcome based education the first requirement is to develop an outcome based curriculum and incorporate an outcome based assessment in the education system. By going through outcome based assessments evaluators will be able to evaluate whether the students have achieved the outlined standard, specific and measurable outcomes. With the proper incorporation of outcome based education there will be a definite commitment to achieve a minimum standard for all learners without giving up at any level. At the end of the program running with the aid of outcome based education, a student will be able to arrive at the following outcomes:

PO1. Engineering knowledge: Apply the knowledge of mathematics, science, engineering fundamentals, and an engineering specialization to the solution of complex engineering problems.

PO2. Problem analysis: Identify, formulate, review research literature, and analyze complex engineering problems reaching substantiated conclusions using first principles of mathematics, natural sciences, and engineering sciences.

PO3. Design / development of solutions: Design solutions for complex engineering problems and design system components or processes that meet the specified needs with appropriate consideration for the public health and safety, and the cultural, societal, and environmental considerations.

PO4. Conduct investigations of complex problems: Use research-based knowledge and research methods including design of experiments, analysis and interpretation of data, and synthesis of the information to provide valid conclusions.

PO5. Modern tool usage: Create, select, and apply appropriate techniques, resources, and modern engineering and IT tools including prediction and modeling to complex engineering activities with an understanding of the limitations.

PO6. The engineer and society: Apply reasoning informed by the contextual knowledge to assess societal, health, safety, legal and cultural issues and the consequent responsibilities relevant to the professional engineering practice.

PO7. Environment and sustainability: Understand the impact of the professional engineering solutions in societal and environmental contexts, and demonstrate the knowledge of, and need for sustainable development.

PO8. Ethics: Apply ethical principles and commit to professional ethics and responsibilities and norms of the engineering practice.

PO9. Individual and team work: Function effectively as an individual, and as a member or leader in diverse teams, and in multidisciplinary settings.

PO10. Communication: Communicate effectively on complex engineering activities with the engineering community and with society at large, such as, being able to comprehend and write effective reports and design documentation, make effective presentations, and give and receive clear instructions.

PO11. Project management and finance: Demonstrate knowledge and understanding of the engineering and management principles and apply these to one's own work, as a member and leader in a team, to manage projects and in multidisciplinary environments.

PO12. Life-long learning: Recognize the need for, and have the preparation and ability to engage in independent and life-long learning in the broadest context of technological change.

Course Outcomes

After completion of the course the students will be able to:

CO-1: Identify crystal structures for various materials and one could understand the defects in such structures.

CO-2: Tailor material properties of ferrous and non-ferrous alloys?

CO-3: Quantify mechanical integrity and failure in materials?

Course Outcomes	Expected Mapping with Programme Outcomes *(1- Weak Correlation; 2- Medium correlation; 3- Strong Correlation)*											
	PO-1	PO-2	PO-3	PO-4	PO-5	PO-6	PO-7	PO-8	PO-9	PO-10	PO-11	PO-12
CO-1	3	2	3	2	1	-	-	-	-	-	-	-
CO-2	2	2	3	2	2	-	-	-	-	-	-	-
CO-3	2	2	3	3	2	-	-	-	-	-	-	-

Guidelines for Teachers

To implement Outcome Based Education (OBE) knowledge level and skill set of the students should be enhanced. Teachers should take a major responsibility for the proper implementation of OBE. Some of the responsibilities (not limited to) for the teachers in OBE system may be as follows:

- Within reasonable constraint, they should manoeuvre time to the best advantage of all students.
- They should assess the students only upon certain defined criterion without considering any other potential ineligibility to discriminate them.
- They should try to grow the learning abilities of the students to a certain level before they leave the institute.
- They should try to ensure that all the students are equipped with the quality knowledge as well as competence after they finish their education.
- They should always encourage the students to develop their ultimate performance capabilities.
- They should facilitate and encourage group work and team work to consolidate newer approach.
- They should follow Blooms taxonomy in every part of the assessment.

Bloom's Taxonomy

Level	Teacher should Check	Student should be able to	Possible Mode of Assessment
Create	Students ability to create	Design or Create	Mini project
Evaluate	Students ability to justify	Argue or Defend	Assignment
Analyse	Students ability to distinguish	Differentiate or Distinguish	Project/Lab Methodology
Apply	Students ability to use information	Operate or Demonstrate	Technical Presentation/ Demonstration
Understand	Students ability to explain the ideas	Explain or Classify	Presentation/Seminar
Remember	Students ability to recall (or remember)	Define or Recall	Quiz

Guidelines for Students

Students should take equal responsibility for implementing the OBE. Some of the responsibilities (not limited to) for the students in OBE system are as follows:

- Students should be well aware of each UO before the start of a unit in each and every course.
- Students should be well aware of each CO before the start of the course.
- Students should be well aware of each PO before the start of the programme.
- Students should think critically and reasonably with proper reflection and action.
- Learning of the students should be connected and integrated with practical and real life consequences.
- Students should be well aware of their competency at every level of OBE.

Abbreviations

General Terms			
Abbreviations	**Full form**	**Abbreviations**	**Full form**
AISI	American iron and steel institute	CE	carbon equivalent
APF	Atomic packing factor	Fe-C	Iron-carbon
BCC	Body-centered cubic	GP zone	Guinier Preston zone
BCT	Body-centered tetragonal	HCP	Hhexagon close-packed unit cell
BHN	Brinell hardness number	HRB	Rockwell hardness on the B scale
C	Centigrade	HRC	Rockwell hardness on the C scale
CCP	Cubicle closed packed	HKN	Knoop hardness number
CCE	Compensated carbon equivalent	Ni eq	Nickel equivalent
CCT	Continuous cooling transformation	TTT	Temperature time transformation
Creq	Chromium equivalent	VHN	Vickers hardness number

List of Figures

CONTENTS

Unit 1

Fundamentals of Material Science

Crystal Structure, Deformation Mechanisms and Strengthening of Metals

Unit Specific / Learning Objective

Objective this unit in to talk about following aspects

- To introduce common engineering materials
- To learning about bonding mechanism at atomic / molecular scale
- To introduce importance crystal structure and its effect materials characteristics
- To determine atomic packing factors of different crystal structures
- To learn about atomic scale deformation mechanisms namely slip and twinning
- Understand the importance of critical resolved shear stress on strength of polycrystalline materials
- To understand through dislocation theory why real crystals weaker than perfect crystals
- To introduce metal strength mechanism to modify the metal characteristics

Additionally, few fundamental questions for self-assessment based on fundamentals **are** included in this chapter in form of application, comprehension, analysis and synthesis. There are further suggested readings and reference for deep learners and readers assistance.

Rationale

Today's human lives are surrounded by variety of materials as these are closely integrated in our daily routine to make our lives easier and comfortable. These (materials) can be found in the form of furniture, automotive, accessories, kitchen etc. The performance and reliability of these items significantly depends on the materials and their atomic, micro and macro scale constituents. Learning about atomic, molecular and their fine particles arrangement in materials helps us to understand, how can physical and mechanical characteristics of materials get affected / modified. Understanding of crystal structure, deformation mechanisms and metal strengthening mechanism will allow young minds to think independently and innovatively to engineer the new materials and therefore, modify the existing one.

Pre-Requisites (To be incorporated by Author as per sample below)

Mathematics: Trigonometry (Class XII)

Chemistry: Inorganic materials (Class XII)

Learning outcomes

U1-O1: Ability to identify different type of crystal structures and crystalline defects of common metals

U1-O2: Ability to develop approach to improve the mechanical properties using suitable metal strengthening mechanism

U1-O3: Ability to interpret and estimate physical and mechanical characteristics of metals based on crystal structure

U1-O4: Ability to calculate the atomic packing factor of common crystal structures

Unit-1 Outcomes	EXPECTED MAPPING WITH COURSE OUTCOMES (1- Weak Correlation; 2- Medium correlation; 3- Strong Correlation)					
	CO-1	CO-2	CO-3	CO-4	CO-5	CO-6
U1-O1	3	3	1			
U1-O2	2	3	1			
U1-O3	2	3	1			
U1-O4	3	1	3			
U1-O5						

Course Objective

1. Student will be able to identify crystal structures for various materials and one could understand the defects in such structures.
2. Understand how to tailor material properties of ferrous and non-ferrous alloys?
3. How to quantify mechanical integrity and failure in materials?

1.1 Materials in society

Materials have always played a major role in civilisation from the Stone Age to the Modern material age. During the ancient times, varieties of materials were primary used to develop weapons and tools for hunting purposes. The requirement of developing effective strong tools led to the development of materials in the form of stone, bronze, copper, tin, and later iron. Accordingly, these were termed as the Stone Age, Bronze Age, Iron Age, and Modern material age (including concrete, polymers, semiconductors, composites, and ceramics). The relative importance of these materials (metals, polymers, ceramics, and composites) has been changing with time as shown in Fig. 1.1.

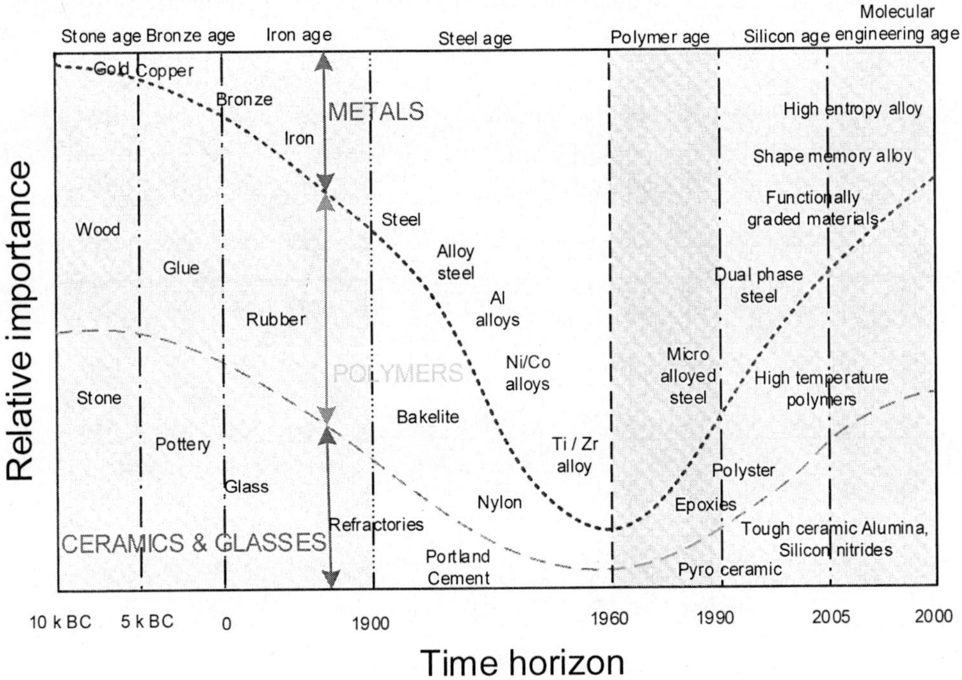

Figure 1.1 Schematic diagram showing varying relative importance of different materials as a function of time

1.2 Science and Engineering of Materials

These materials are made of atoms with different electronic structures, and molecules. These constituents (of varying scale from sub-atomic, atomic, crystal structure and microstructure) are found to be arranged and bonded in different ways. The arrangement of atoms and molecules and their bonding between them determines their properties and service performance. It is therefore important to understand and learn more about the arrangement and bonding between atoms, molecules and compounds.

Material science is basically an interdisciplinary domain combining solid-state physics, process engineering, chemistry, mathematics, mechanical engineering, mechanics, management, applied computer science, biology, and medicines to make products for the society so as to improve quality, reliability, and life, and the availability to all. The aim of material science is to study the effect of the structure of the material on the different scales ranging from electrons, crystals, and micro and macro-level on the characteristics of the material. Proper understanding of material science led to the development of many innovative and normal products in the form of ceramics, exotic materials, nanomaterials, semiconductors, and polymers.

On the other hand, materials engineering basically focuses on the relationship between the structure and properties, tailoring properties of materials, and structural modifications using appropriate manufacturing processes. Structural modification can be obtained by changing the crystal structure and microstructure of the materials to achieve the desired set of functional properties such as mechanical properties, tribological properties, physical properties, etc. Therefore, materials engineering helps in development of the products for improved performance and reliability (Fig. 1.2).

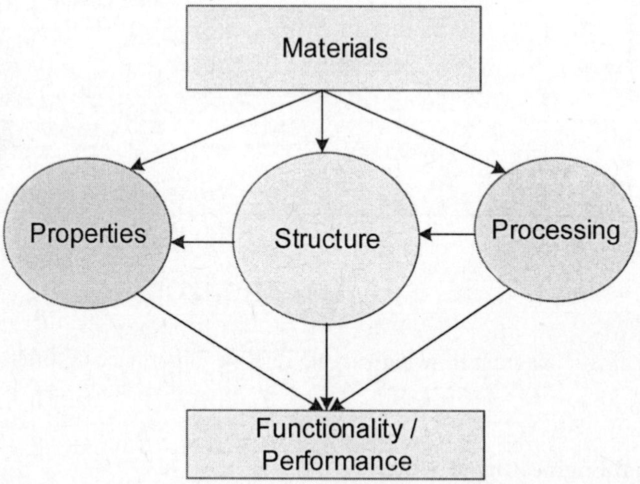

Figure 1.2 Schematic diagram showing inter-relationship between materials, manufacturing and properties

1.3 Relevance of Material Engineering

A good understanding of materials engineering helps in the selection of suitable manufacturing processes to develop the desired microstructure in the final products to achieve the target functional properties for the improved performance. This, in turn, reduces the chances of failure during the service and improves reliability. Therefore, the selection of suitable materials at the design stage before the manufacturing stage, becomes very crucial. The different materials respond in different ways to a given manufacturing process due to varying thermal and mechanical loading. Further, different materials need different manufacturing processes to make the product more efficiently and economically. Materials engineering helps to estimate and anticipate the ways by which the manufacturing process will affect the structure and so their functional properties.

For the design purpose, it is very important to consider the functional properties desired and the availability of the material, the cost and the processing method needed for manufacturing. Further, the choice of the processing method is strongly influenced by component geometry including size and shape, the properties of the material to be processed, the target properties as per the design, dimensional accuracy and finish desired, and availability of the manufacturing setup, and finally the cost (Fig. 1.3).

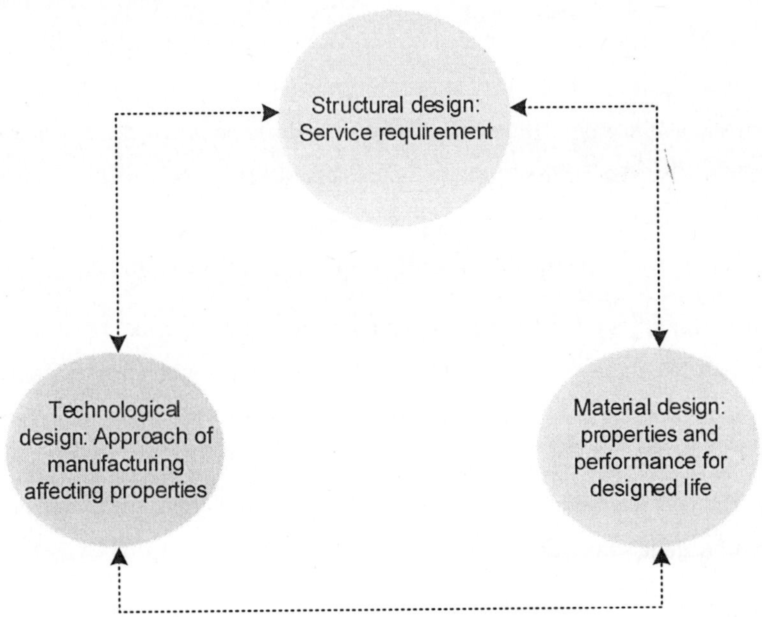

Figure 1.3 Schematic diagram showing inter-relationship between structural, materials and technological designs

1.4 Scope of Materials Engineering

The scope of materials engineering includes physical metallurgy and mechanical metallurgy. Physical metallurgy is about the structure and physical properties of the metals, along with the mechanism affecting the properties. Additionally, physical metallurgy comprises mechanical property evaluation, heat treatment, and metallography. Mechanical property evaluation involves the assessment of metals under external loading conditions. Mechanical properties such as hardness, toughness, tensile strength, ductility, fracture toughness, fatigue resistance, and creep resistance as per the application requirements are evaluated. Heat treatment of the metals primarily involves the imposition of the controlled thermal cycle of heating, and soaking, followed by cooling at different rates to achieve the desired microstructure and mechanical properties. Metallography, on the other hand, is about investigating the phase and grain structure of the metals using suitable tools like optical and electron microscopes and X-ray diffraction analysis to establish the relationship between the structure and other properties of the material. Mechanical metallurgy includes all thermomechanical processes used to achieve the desired shape of metals and alloys, e.g. casting, welding, forming, drawing etc.

1.5 Common Engineering Materials

In our daily life, everyone is surrounded by various products that make our life easier. These products are made of a wide range of materials from wood, glasses, concrete, rubber, plastics, steel, alumina, and silicon. To understand these products as an engineer; materials can be grouped into five categories: metals, ceramics, glasses, polymers, composites, and semiconductors. Out of these five materials, the first three (metals, ceramics & glasses, polymers) are commonly used to design and manufacture various products.

Most of the composites are developed by combining any of two (or more) amongst these first three materials like metal-polymer, metal-ceramics, and ceramic-polymers. Composites offer certain properties which are far better than those of individual materials. In composites, each constituent retain its identity. The fifth material semiconductors, is primarily used in the electronic and computer industries. The following section highlights the significance of these five types of materials in engineering.

1.5.1 Metals

Metal is the most important materials found in every facet of life. Metals show perfect combination of mechanical properties such as tensile strength, hardness, toughness, ductility, and resistance to fracture. Additionally, most metals are good electrical and thermal conductors. A combination of low yield strength and high ductility of metals facilitates the permanent plastic deformation on applying the stresses above the yield strength. This feature of metals allows manufacturing of metals (to make products) using plastic deformation approach such as forming. The most common metals of great commercial importance include steel, cast iron, aluminium alloys, copper and copper alloys, magnesium alloys, nickel alloys, titanium alloys, zinc alloys etc. The different metals offer a very wide range of mechanical, thermal, and physical properties. For each type of metal, there are multiple alloys offering a wide range of mechanical, chemical, metallurgical, tribological and physical properties. For example, steel as per their composition, crystal structure, and microstructure, offer strength in the range from 400 MPa to 2000 MPa, with ductility varying from 10-80 % and impact toughness from 10 J to 200 J, and fracture toughness from 40 MPa to 200 MPa m1/2, and hardness from 5 HRC to 60 HRC. The role of such composition is somewhat limited compared to the crystal structure and microstructure.

1.5.2 Ceramics

Ceramic and glasses are another essential categories the materials. Ceramics are compounds of one / more metallic elements reacting with the five non-metallic elements such as carbon, nitrogen, oxygen, phosphorus, and sulphur. The most common ceramic materials are alumina, silicon carbide, tungsten carbide, and silicon nitride. Ceramic materials are very stable at high temperatures and in harsh chemical environments. High-temperature stability of the ceramics leads to increased resistance to thermal softening. These two characteristics make the ceramic materials suitable for cutting and grinding applications. However, ceramic materials suffer from high brittleness, poor ductility, and low toughness. Therefore, ceramic materials do not find many applications in the manufacturing of structural components. Recent development in ceramic technology has resulted in newer fracture tough ceramic materials like silicon nitride for applications in energy-efficient jet engine components. Most ceramic materials like metals are crystalline in terms of atomic arrangement. However, these can be developed made of non-crystalline structures as well. In crystalline materials, atoms are arranged in regular and ordered manner with repeating pattern. In contrast, non-crystalline materials have an atomic arrangement that is regular, disordered, and follows non-random patterns.

1.5.3 Polymers

Polymer is a modern material also known as plastic. These are synthetic materials as products of organic chemistry. Polymer is composed of two words: poly means multiple and mer stands for single hydrocarbon molecules such as ethylene, silicones, fluorine, nitrogen, and oxygen. These are composed of long-chain molecules. Polyethylene means many (a few hundred to thousands) ethylene molecules bonded together as a long chain. These long-molecular chains of polymers can be weakly or strongly bonded together, which determines their responses to thermal and mechanical loading. There are two board categories of polymers, namely thermoplastic and thermosetting. Thermosetting polymers are composed of long-chain molecules which are strongly bonded and cross-linked. Thermosetting plastics are stronger and more stable than thermoplastics. Therefore, thermosetting polymers find few applications in manufacturing structural components, especially in the automotive sector. In general, the polymers are known to be light in weight, low in strength and thermal stability, and high in chemical sensitivity. Efforts are being made to develop biodegradable, high-strength polymers for structural application as a possible replacement for structural metals.

1.5.4 Composites

Composite is another interesting and essential category of materials that has received the attention of scientists and technologists. However, there are few natural composite materials like wood having fiber-reinforced structures. The composite materials (mostly designed to deal with the challenges of harsh service conditions) are composed of matrix materials like metals, polymers, and ceramics reinforced with suitable agents in the form of powders, fibers, sheets, films, etc. Composite materials help to realise the desired properties from different kind of materials in a single material (Fig. 1.4). For example, copper reinforced with graphite helps to achieve good thermal conductivity and lubricating properties used for making electric bush used in motors. Similarly, nickel matric reinforced with tungsten carbide helps to achieve high thermal stability, good toughness and high hardness for many high-temperature tribological applications.

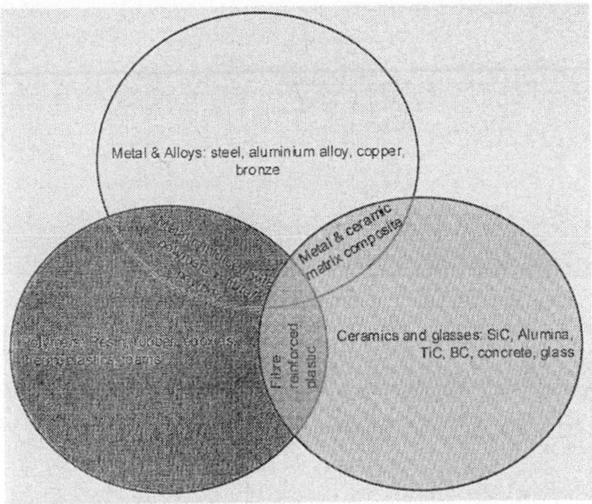

Figure 1.4 Schematic diagram showing the three most common engineering materials and their combinations that can be used to develop composites

1.5.5. Semiconductors

The semiconductor material is most commonly used in the electronic goods (computers, mobile and electronic gadgets) that we use daily, but it is not so easily visible. It is not visible like other materials like polymers, metals, or composite materials in the form of concrete. Semiconductor materials are neither good nor bad conductors. Because of their moderate level of electrical conductivity, these are called semiconductors. Silicon and Germanium are the commonly used semiconductors. The purity of the semiconductor materials determines their electrical behaviour. Precise control of the purity during semiconductor manufacturing is needed to achieve the desired set of electrical properties.

1.6 Crystal structure

Atoms are the building blocks of the matter we see around in the form of solid, liquid, and gas. The way atoms are arranged in a matter, and the binding force between them determines their characteristics and responses to external conditions such as load, heat, electricity, and light. The arrangement of the atoms in the liquid state is somewhat random and follows short-range order. The bonding force between the atoms in the liquid state is very weak. Therefore, they tend to separate and bond again frequently. Materials in the liquid state are characterised as random grouping, scattering, and regrouping for a short period. In the case of the gaseous state, the arrangement of the atoms is extremely random, and bonding between them is very weak; therefore, random grouping is rare while the atoms move easily.

1.7 Bonding Mechanisms

Atoms and molecules are bonded together through many bonding mechanisms. Van der walls force, ionic bonding, covalent bonding, and metallic bonding are the four common bonding mechanisms.

1.7.1 Van der Waals force

Van der waals force results in a weak bond when atoms and molecules exhibit permanent dipoles due to the segregation of electrons on one side of the atoms/molecules (Fig. 1.5). This kind of bonding plays an important role in polymers where long molecular chains have a strong bond between the atoms of the chain. In contrast, the long molecular chains them-self are bonded together through the Van der waals force bonds. Thermoplastics are a typical example of where such kinds of bonds are formed.

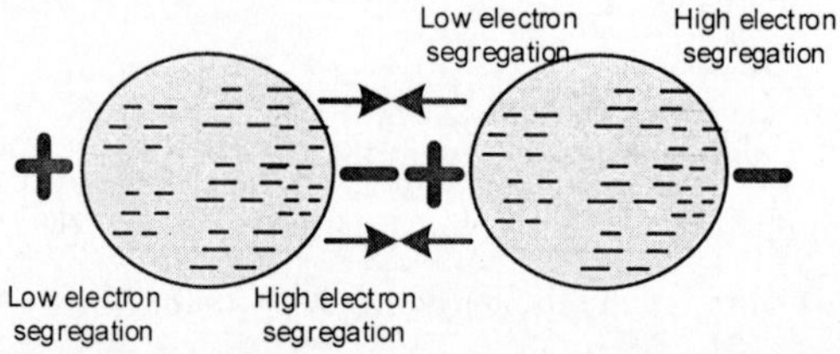

Figure 1.5 Schematic diagram of bonding due to Van der waals force

Ionic, covalent, and metallic bonds form strong joints between the atoms and molecules resulting in many hard, strong, and refractory materials such as diamond, ceramics, and tungsten.

1.7.2 Covalent bond

The covalent bond formation is based on sharing the electrons between the atoms, which helps them attain the stable electron configuration in their outer shell. The sharing of electrons by the atoms affects the properties of the materials. For example, the four electrons shared by a carbon atom with other four atoms result in a tetrahedron arrangement leading to the hardest material as a diamond, while sharing of three electrons by a carbon atom with three other atoms (making fourth electrons free) produces soft graphite form of carbon which is used as a lubricant. The properties of material property significantly depends on the type of covalent bonds formed and their atomic arrangements.

1.7.3 Ionic bond

An ionic bond is formed due to the electrostatic attraction between the oppositely charged ions. The elements like oxygen, chlorine, and sulphur easily receive a few electrons to achieve stability and are termed electronegative elements. The metals, on the other hand, dissociate to provide free electrons and are called electropositive elements. For example, sodium ions form ionic bonds with chlorine. Similarly, many ceramics like magnesia, alumina, and zirconium oxide are formed through the ionic bond formation.

1.7.4 Metallic bond

The metallic bond is formed in metals by forming free-electron cloud around the spaces between the atoms (Fig. 1.6). The force of attraction between the positively charged ions and electron cloud results in the desired coherence/bonding force. The presence of free electrons imparts thermal and electrical conductivity to the metals. The metallic bonding depends on the number of valence electrons associated with atoms of an element. Good metallic bonding is observed when metallic elements have few valence electrons. The fewer the valency of electrons of an element, this results in a higher tendency to make free electrons available for metallic bonding.

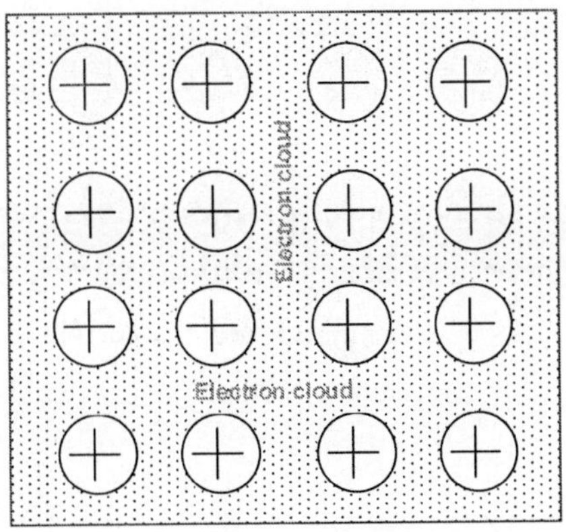

Figure 1.6 Schematic diagram of metallic bonding

The arrangement of the atoms in the case of solid can be in the form ordered array or random. At a low energy level, atoms in a few solids (mostly in metals) form clusters in an ordered manner with long-range order, which are very closely packed, and the bonding force between them is extremely strong. There can be a variety of arrangements in which atoms tend to be clustered. Such arrangements (ordered array of atoms) are repeated in all directions leading to a long-range order. An ordered array of atoms results in a lattice. A lattice shows the way by which atoms are arranged and are being repeated. The typical arrangement of the atoms in a lattice is called lattice structure. The solids (like most metals, ceramics, and a few polymers) have a lattice structure of long-range order called crystalline materials. However, many polymers, a few metals and ceramics even in a solid-state show random-ordered arrangement over a long-range is called non-crystalline or amorphous materials.

1.8 Unit cell of Crystal Structures of Metals

The crystal structure is characterised as regular and repeating in nature. To understand this regular and repeating nature in the three-dimensional space of a crystal structure, it is important to see how the atoms are arranged in a cell which is repeating. Unit cell indicates the smallest group of atoms along with their arrangements, which is repeating in the three-dimensional space of a crystal structure. There are two aspects which need attention a) shape of the unit cell and b) how the atoms are arranged in a unit. There are 7 different unit cell shapes in which atoms can be arranged, such as cubic, tetragonal, orthorhombic etc. These are characterised by the length of sides and their angle. For example, the most common cubical unit cell has an equal length of sides and all are at a right angle. Further, atoms can be arranged in these 7 shapes of the unit cell in total 14 ways called Bravis lattices.

For example, a simple cubical unit cell based on atomic arrange forms four different lattice structures, namely simple cubical, body centre cubical, face centred cubical, and hexagonal closed packed crystal structures (Fig.1.7). The unit cells in a crystalline structure repeat in three dimensional space as an array of atoms arranges in a specific way. The number of atoms associated with each unit needs a careful observation regarding atoms being present at the lattice points, faces and bases of the unit cell are shared with other unit cells in all three directions. This sharing directly affects the effective number of atoms with a unit cell.

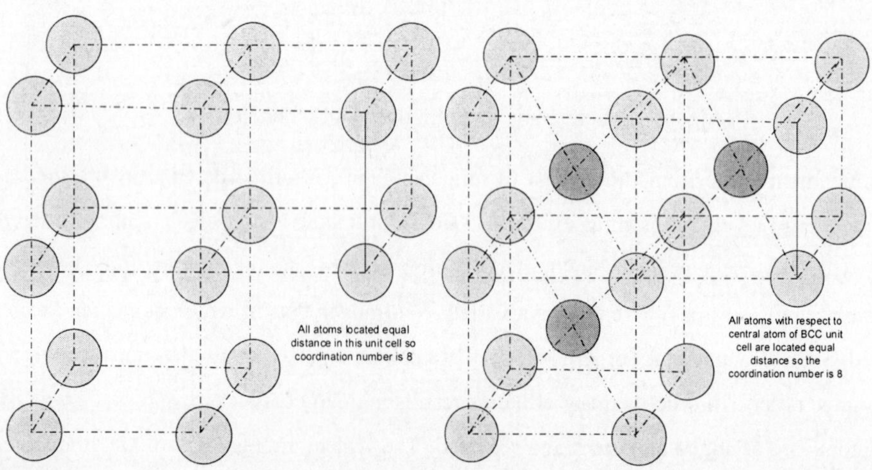

a) b)

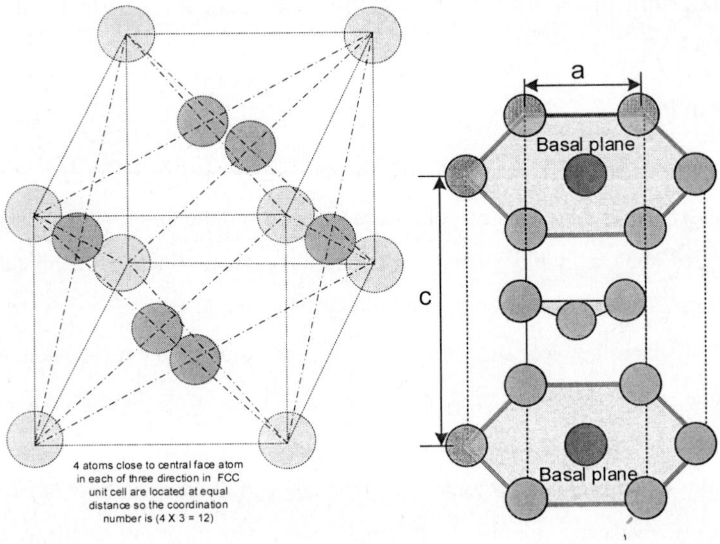

c) d)

Figure 1.7 Schematic diagram of atomic arrangement in unit cells of a) cubical, b) body centred, c) face centred and d) hexagonal packed crystal structure

For example, a cubical unit cell has 8 atoms at eight lattice points, but all of these are shared with 8 adjoining units' cells in three dimensional space. Thus the effective number atom with cubical closed packed unit cell (CCC) is 1 as obtained from 8/8. The body centred cubic (BCC) cell has 8 atoms at eight corners lattice points and one atom at the centre of the unit cell, which is not shared with any other unit cell. Therefore, the effective number of atoms with BCC is 2 obtained from 8/8 + 1. Similarly, face centred cubic (FCC) cell has 8 atoms at eight lattice points and one atom at each of 6 faces of the unit cell, which is shared with other adjacent unit cells. Therefore, the effective number of atoms with FCC is 4 obtained from 8/8 + 6/2. Each atom at the centre of six faces shared with adjoining unit cell. Thus, effectively only half of an atom on each face is associated with a cell unit. The number of equal distance atoms in the vicinity of central atoms of a unit cell in three-dimensional space is called coordination of a unit cell. For example, for FCC, the coordination number of atoms is 12 (4 corner atoms at equal distance from central atoms on each face on each of three directions of three dimensional space leading to 4 X 3). For BCC, there are 8 corner atoms at an equal distance from the central atom, so 8 is the coordination number. Similar for CCC, the coordination number is 6. The effective number of unit cell of a lattice structure directly affects the atomic packing factor

(AFP) also called packing efficiency. The AFP for a crystal structure is obtained from ratio of volume of atoms in a unit cell to the volume of unit cell itself. Assuming the spherical shape of atoms each of radius R.

1.9 Effect of crystal structure

The crystal structure as per the arrangement of atoms in a given shape of the unit cell determines the physical properties like density, solid-state solubility to the alloying elements and mechanical properties such as yield strength and ductility. The influence of crystal structure on mechanical properties is governed by the following factors:

- The availability of spaces between atoms in a unit cell for accommodating the solute atoms affects solid solubility. An increase in solid solubility enhances the mechanical properties by a solid solution strengthening mechanism.
- The number of slip systems that are present in a given unit cell. A large number of slip system reduces the resistance to plastic deformation. As deformation mechanism becomes active at a much lower level of stresses in metal having more slip systems (copper and aluminium) than those having less (Fe, W) number of slip systems.

1.10 Atomic Packing Factor

The atomic packing factor indicates how closely atoms are packed in a unit cell, and it is obtained from the ratio of volume of all (effective) atoms to volume of the unit cell itself. Since the effective number of atoms depends on the crystal structure of the material, therefore atomic packing factor also varies accordingly.

1.10.1 Cubic Closed Packed (CCP)

This crystal structure has a cubical unit cell with eight atoms at all corner lattice points of the units (Fig. 1.8).

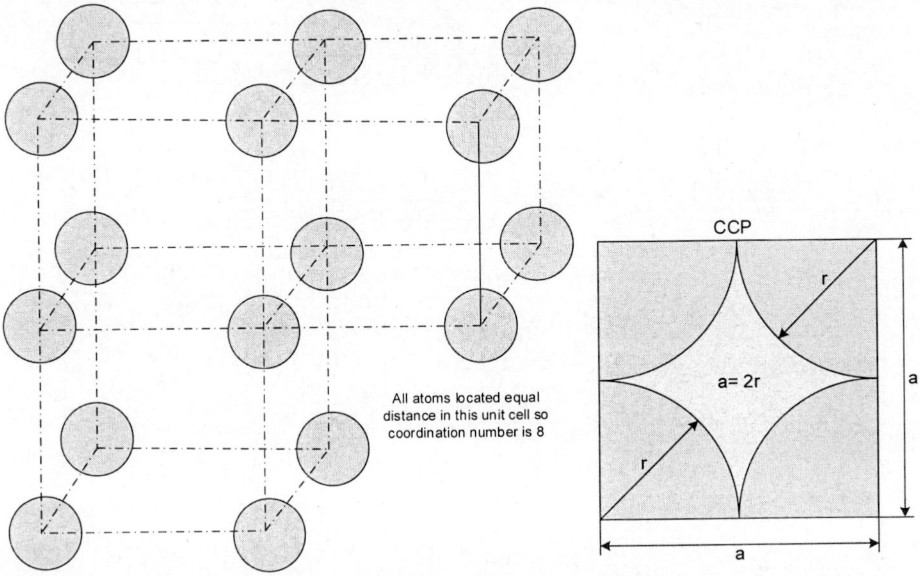

a) b)

Figure 1.8 Schematic diagram of a) atomic arrangement in unit cells of cubical closed packed crystal structure and b) visualisation of atomic packing on one side of a cubic cell

Coordination number of atoms for CCP unit cell: 8

Effective number of atoms in CCP: 8/8=1

Atomic packing factor: volume of atoms in a unit cell / volume of unit cell = effective number of atoms X volume of one atom / volume of the unit cell

Side of the cubical cell a is equal to 2r

Volume of unit cell a^3 or $(2r)^3$

Volume of an atom: $4\pi (r)^3/3$

Atomic packing factor: $[1 \times 4\pi (r)^3/3]/(2r)^3 = \pi/2\times3$: 0.52

1.10.2 Body Centred Cubic

This crystal structure has a cubical unit cell with eight atoms at all corners lattice point and one atom at centre of the units (Fig. 1.9).

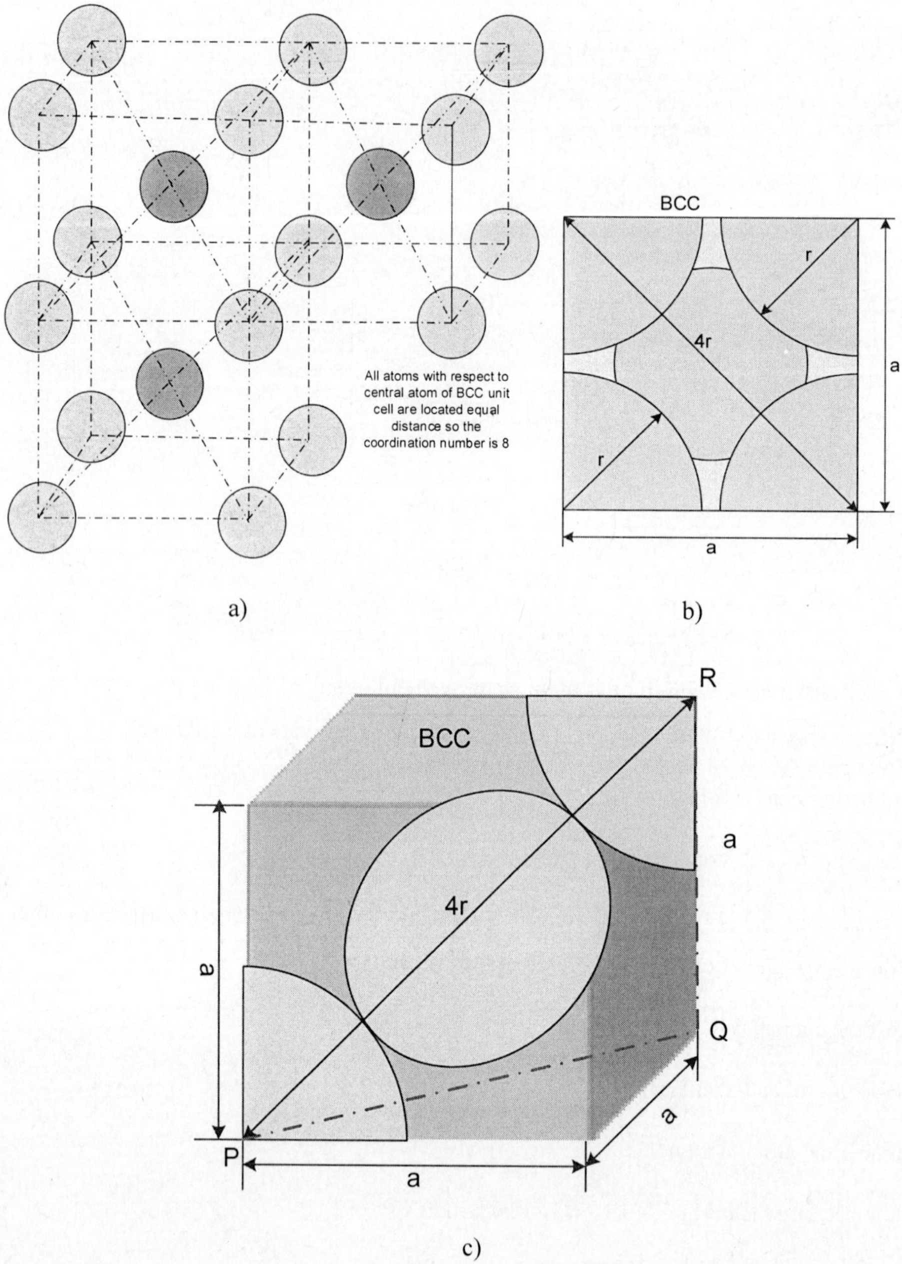

Figure 1.9 Schematic diagram of a) atomic arrangement in unit cells of body centred cubic crystal structure, b) visualisation of atomic packing from one face of a body centred cubic cell and c) 3-D view of atomic arrangement along the diagonal of body centred cubic unit cell

Coordination number of atoms for BCC unit cell: 8

Effective number of atoms in BCC crystal structure: 2

Atomic packing factor: volume of atoms in a unit cell / volume of unit cell = effective number of atoms X volume of one atom / volume of unit cell

Side of cubical cell a. Diagonal of unit cell PR is equal to 4r. $PR = (PQ^2 + QR^2)^{1/2}$ where $PQ = (a^2 + a^2)^{1/2} = (2a^2)^{1/2}$ and QR is a. Therefore, $PR = 4r = [\{(2a^2)^{1/2}\}^2 + a^2]^{1/2} = [\{(3a^2]^{1/2}$

Solving further (by obtaining square of both sides) gives us $16r^2 = 3a^2$ so $a = (16r^2/3)^{1/2} = 4r/3^{1/2}$

Volume of unit cell a^3 or $(4r/3^{1/2})^3$

Volume of an atom: $4\pi(r)^3/3$

Atomic packing factor: $[2 X 4\pi(r)^3/3]/(4r/3^{1/2})^3 = 0.68$

1.10.3 Face Centred Cubic

This crystal structure has a cubical unit cell having eight atoms at all corner lattice point and 6 atoms at the centre of each of the six faces of the unit cell (Fig. 1.10)

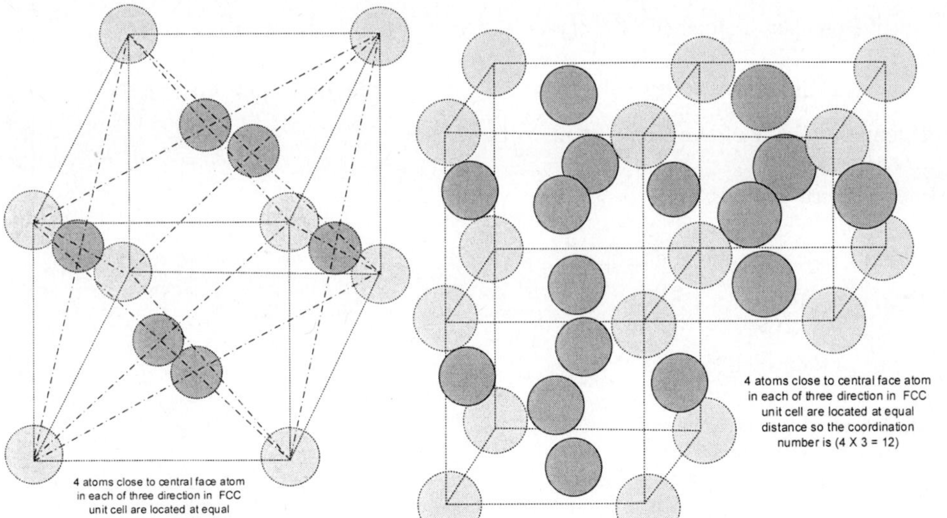

a) b)

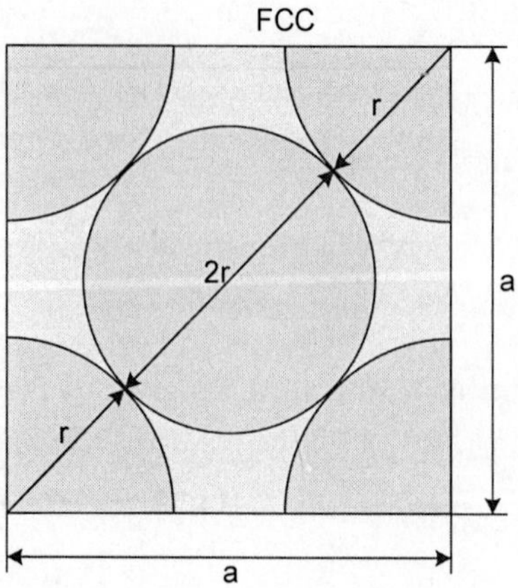

c)

Figure 1.10 Schematic diagram of a) atomic arrangement in unit cells of face centred cubic crystal structure, b) continuity of atomic arrangement in face centred cubic crystal structure, and c) 2-D view of atomic arrangement along from one face of a face centred cubic cell

Coordination number of atoms for FCC unit cell: 12

Effective number of atoms in FCC crystal structure: 4

Atomic packing factor: volume of atoms in a unit cell / volume of unit cell = effective number of atoms X volume of one atom / volume of unit cell

Side of cubical cell a is equal to $4r = (a^2 + a^2)^{1/2} = (2a^2)^{1/2}$.

Solving further (by obtaining square of both sides) gives us $16r^2 = 2a^2$ so $a = (16r^2/2)^{1/2} = (8r^2)^{1/2} = 2r(2)^{1/2}$

Volume of unit cell a^3 or $(2r\ X\ 2^{1/2})^3$

Volume of an atom: $4\ \pi\ (r)^3/3$

Atomic packing factor: $[4\ X\ 4\ \pi\ (r)^3/3]/\ (2r\ X\ 2^{1/2})^3$=0.74

1.10.4 Hexagonal Closed Packed

The crystal structure has two hexagonal shape basal planes with 12 atoms at corner lattice points along with three atoms at the middle level and atoms basal planes (Fig. 1.11).

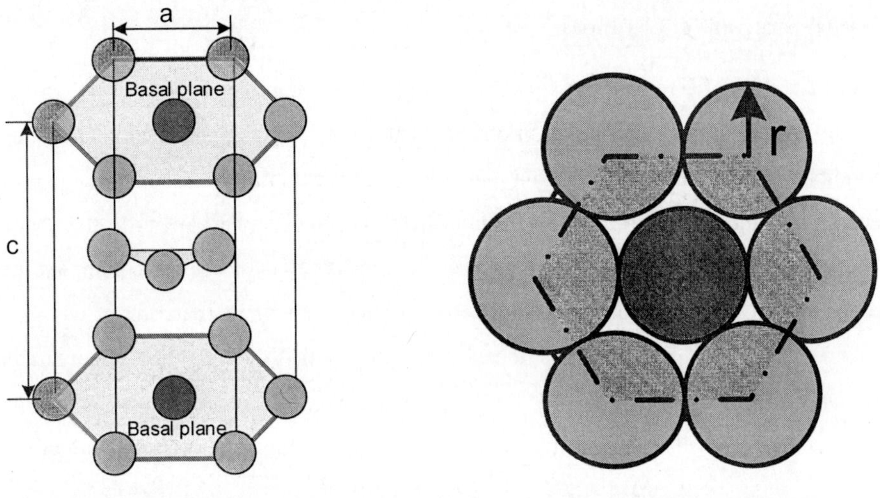

a) b)

Figure 1.11 Schematic diagram of a) atomic arrangement in the unit cell of hexagonal closed packed crystal structure, and b) atomic arrangement at the basal plane hexagonal closed packed crystal structure

Coordination number of atoms for HCP unit cell: 12

Effective number of atoms in HCP crystal structure: 6 (2+ 3+ 1) [a) 1/6 of 12 each corner atoms 12/6 : 2, b) 3 atoms within cell 3/1 : 3, and c) ½ of 1 atom each at basal plane within unit cell (2/2: 1)]

Atomic packing factor: volume of atoms in a unit cell / volume of unit cell = effective number of atoms X volume of one atom / volume of the unit cell

Side of unit cell $a= 2r$ and height of unit cell $c = 2\ (2/3)^{1/2}\ a$

Volume of unit cell: area of hexagon X height of unit cell: $[3(3)^{1/2}\ a^2]/2\ X\ 2\ (2/3)^{1/2}\ a\ :\ 24\ (2)^{1/2}\ r^3$

Volume of an atom: $4\ \pi\ (r)^3/3$

Volume of effective atoms of unit cell: $6\ X\ 4\ \pi\ (r)^3/3 = 8\pi\ (r)^3$

Atomic packing factor: $[(8\pi\ (r)^3)/24\ (2)^{1/2}\ r^3 = 0.74$

The atomic packing factors of HCP crystal structure higher than FCC, BCC and CCP

1.11 Crystal Structure of Ceramics

The crystal structure of ceramics can vary from simple to very complex. These can be entirely amorphous, crystalline, or a combination of both. The crystalline zone is usually surrounded by amorphous in the latter case. Ceramic materials usually exhibit ionic bonding along with covalent where cations (metal) and anions (non-metal like oxygen, carbon) form the desired bond (Fig. 1.12). The crystal structure of ceramic materials is found to be more complex than the metals. The presence of strong ionic and covalent bonds in ceramic materials as compared to metals results in very high hardness, tensile strength, rigidity, and stability at temperature against harsh chemical environments. Further, the absence of free electrons in ceramic materials makes them behave like insulators with reduced thermal and electrical conductivity.

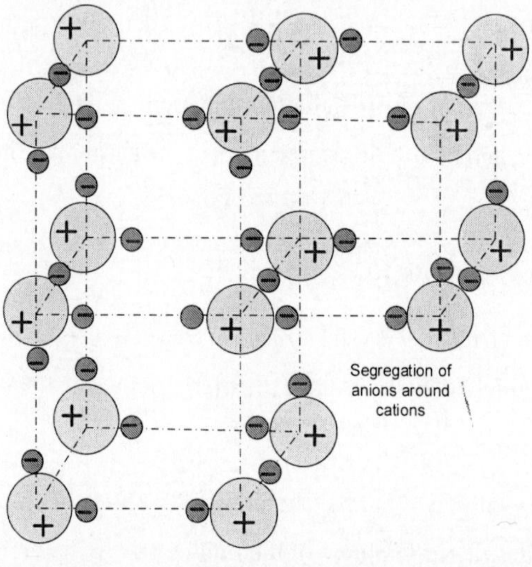

Figure 1.12 Schematic diagram of the crystal structure of ceramic materials with ionic bonding

Cations being significantly larger than the anions are surrounded by many anions. The cations are positively charged metallic ions, while anions are negatively charged particles. These are usually composed of two or more elements for example, alumina is a compound of aluminium and oxygen, and similarly, silicon carbide and tungsten carbide are the compounds of carbon with silicon and tungsten, respectively.

Ceramic materials are found to have a very wide range of crystal structures and microstructures that can be tailored through controlled processing and thermal treatments. Therefore, it is possible to get ceramic materials in the form of certain oxides having high thermal insulation capabilities to ceramic materials like silicon carbide and diamond have much higher thermal conductivity than metals like aluminium and copper. Similarly, by controlling the microstructure and crystal structure, it is possible to achieve fracture-tough ceramic materials for structural applications at high temperatures.

1.12 Crystal Structure of Polymers

Compared to the metals and ceramics (where stacking of atoms and iron is observed in two regular and repeating patterns), polymers primarily exhibit an arrangement of the molecules in the form of long chains showing mostly irregular and nonrepeating patterns (Fig. 1.13). Therefore, polymers mostly show a high degree of non-crystallinity. Polymers sometimes show short-range crystalline structures, which are most complex (Fi. 1.14). The short-range crystallinity in polymers usually varies from 20-50 % and is expected to affect the strength, hardness, and density of the polymers.

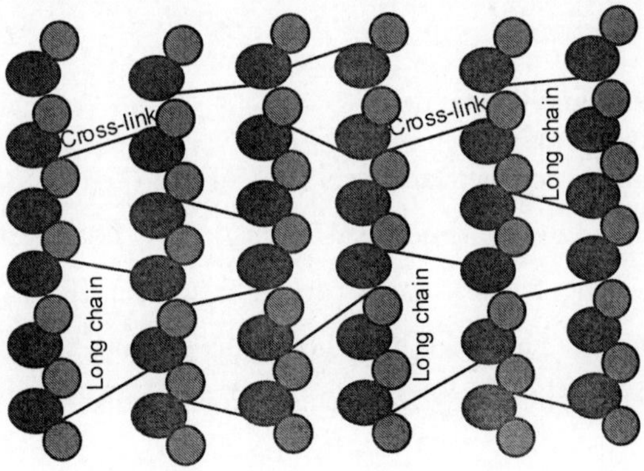

Figure 1.13 Schematic diagram of the crystal structure of polymers with long chain and little cross-linking

Crystallinity indicates the extent of long-range order that exists in polymers. An increase of crystalline results increased regularly aligned molecular chains increasing hardness and density. A common polymer called polyethylene shows the orthorhombic unit cell. The polymers mostly have long-chain molecules (a few hundred nano-meters) arranged in a folding manner back and forth on themselves (on an atomic scale). High-density polyethylene (HDPE) a very common polymer, comprises linear chains with fewer branching. The highly packed molecular structure of HDPE leads to high strength, therefore, suits for manufacturing drainpipes and many other applications. On the other hand low-density polyethylene has many short branches obstruct the close packing of molecules to have an ordered structure leading to reduced strength. Therefore, LDPE suits for making plastic carry bags.

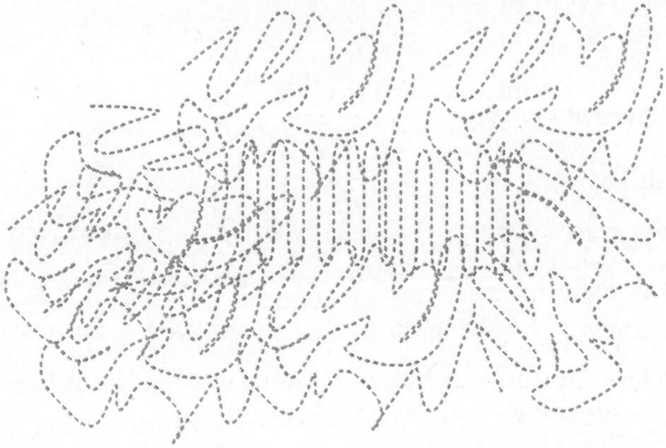

Figure 1.14 Schematic diagram of the crystal structure of polymer with crystalline and amorphous zones

1.13 Crystalline metals and crystal defects

The most common commercial metals show three types of crystal structures, namely, body centered crystal (B. C. C.), face centered crystal (F. C. C.), and hexagonal packed crystals (H.C.P.). The arrangement of atoms, atomic packing factor, the effective number of atoms and their coordination numbers have been described in the previous sections. These aspects of a crystal structure affect the physical and mechanical properties like yield strength, and hardness of metals

The yield strength of a metal determines resistance to plastic deformation, which usually occurs by two mechanisms, i.e. slip, twinning, or a combination of these two. These atomic scale mechanisms facilitate plastic deformation by sliding atoms one over another under the influence of external stresses. The sliding of atoms in metals takes place on certain crystallographic planes. These crystallographic planes (known to facilitate the slip / twinning) possess the maximum atomic density are called slip planes. The sliding of atoms or slip on a slip plane as per external stress and the metal system can take place in many directions. The most closed packed (atoms) direction in a crystal structure called slip direction. Combinations formed by various slip planes and slip directions in a crystal structure facilitating the slip are called slip systems. For example, FCC metal (Aluminium) having 4 slip planes and 3 slip directions shows (4 X 3) 12 slip systems.

Under the influence of external stress, a metal having higher number of slip systems experiences slip, leading to the plastic deformation at lower shear stress. Therefore, plastic deformation of metals having FCC crystal structure (Cu, Al, Au and Ag) takes places easily as compared to metals having BCC (alpha-Fe, W, Cr, V) or HCP (Zn, Mg) crystal structure primarily due to the difference in the number of slip systems.

Further, the distortion in regular crystal structures of metals affects the mechanical properties. For example, transformation of body cubic centered (BCC) body cubic tetragonal crystal structure due to super-saturation of carbon in α-Fe (BCC) leads to the martensitic transformation (BCT). This kind of BCC to BCT transformation causes significant increases in the hardness of iron. An increased degree of distortion in the lattice of crystal structure (due to super-saturation) increases the hardness and strength of metals like α-iron.

Crystal structure of common metals

Sr. No.	Crystal structure	Metals	Crystal structure
1	B. C. C.	α-Fe, W, Co, V, Cr	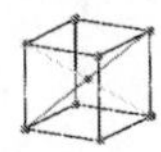
2	F. C. C.	Al, Au, Ag, Mn, Ni, γ-Fe, Cu	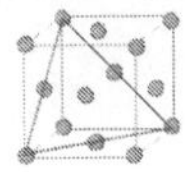
3	H. C. P.	Mg, Zn	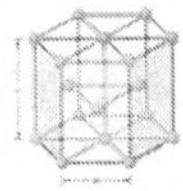

1.14 Crystalline defects

Crystalline metals exhibit many defects like point defects (vacancy, interstitial atoms), line defects (edge / screw dislocation), surface defects (grain boundary, twinning boundaries, stacking fault) and volume defects (twins, precipitates, cracks, pores, inclusions). All these defects (except volume defect) cause lattice strain around the region wherever these are present. The strained lattice increases resistance to deformation.

The point defect in crystalline structure occurs when either regular atom(s) of a crystal is/are missing or foreign atom(s) as an impurity or alloying elements occupy interstitial/substitutional space(s) in a regular arrangement of atoms accordingly. The missing atom crystalline defect called vacancy (Fig. 1.15).

The line defect in crystalline structure also called dislocation occurs when there is anomaly in the arrangement of row of atoms in crystal. This may look like incomplete row of atoms (Fig.1.15). The solidification / plastic deformation of metal results in line defects. The orientation of dislocation can be localized in straight line or spiral form and accordingly these are called edge dislocation and screw dislocation respectively. Screw dislocations occur due to tearing action under the influence of external stress. The presence of dislocation in crystalline metals can improve or deteriorate the mechanical properties.
Surface defects are found in different forms like stacking fault, random arrange of atoms at the grain boundary and twinning boundaries (Fig. 1.16).

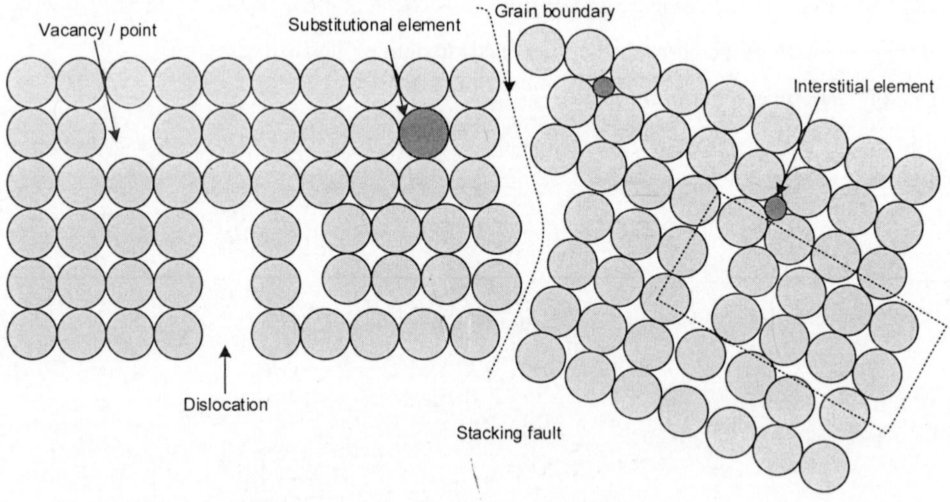

Figure 1.15 Schematic diagram showing variety of crystalline defects like vacancy, dislocation, grain boundaries

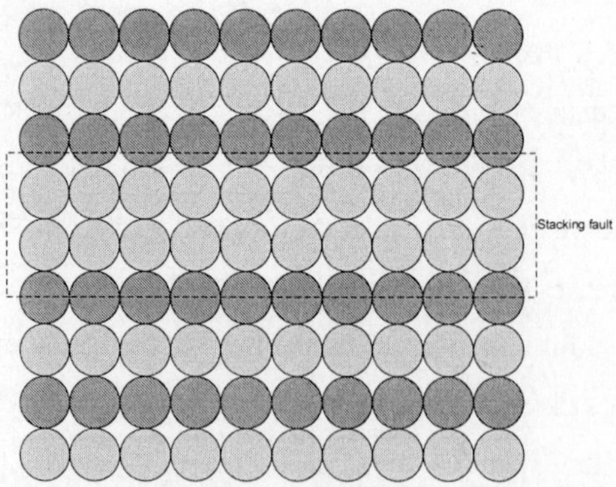

a)

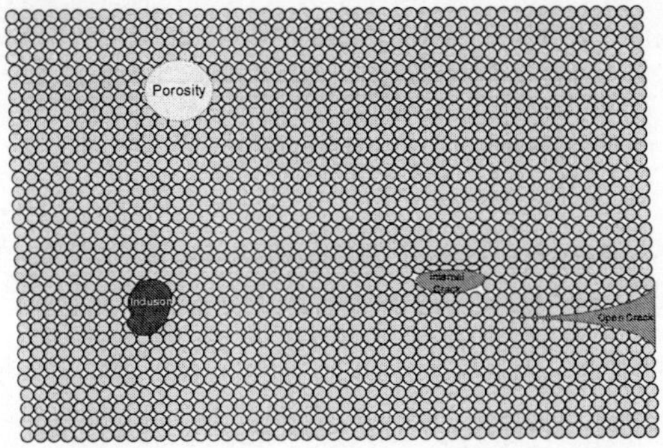

b)

Figure 1.16 Schematic diagram showing a) stacking fault and b) volume crystalline defects

Surface defects like grain boundaries, and stacking fault regions have a dis-ordered arrangement of atoms, which mainly act as major obstacles for moving dislocations. The dis-ordered atomic arrangement at grain boundaries stops the further movement of dislocation therefore presence of more grain boundaries achieved through grain refinement of metals help to improve the mechanical properties. The grain boundaries in the metal can be good or bad from a metal strength point of view. The role of grain boundaries on yield strength is significantly determined by the temperature of metal. A critical temperature associated with each metal called homologous temperature is a demarcating temperature across which the role of grain boundaries on the yield strength of metal is reversed. Grain refinement (increasing grain boundaries) strengthen the metals below the homologous temperature only while grain refinement weakens the metal at above homologous temperature.

Stacking fault is another important crystalline surface defect affecting the response of metals to the plastic deformation. Energy associated with stacking faults, usually termed as stacking fault energy one of the common way to estimate / anticipate response of metals to plastic deformation. Low stacking fault energy metals like high manganese steel, austenitic stainless steel, and cobalt alloy work / stain harden rapidly and increase the stress required for further plastic deformation compared to high stacking fault energy metals like Al, Mg.

Volume defects like cracks, pores, inclusion etc., reduce the strength, load carrying capacity and resistance of metals to static and dynamic fluctuating loads very badly. The reduction in mechanical strength to the static and dynamic loads presence of volume defects such as cracks, pores, inclusions etc. occurs due to two main factors:

a) Defects break the metallic continuity and reduce the load resisting cross sectional area (thereby increasing the actual stresses), which in turn reduces the load carrying capacity of metals

b) Defect act as stress raisers and increase the stress concentration significantly, which in turn triggers the fracture at lower loads.

1.15 Deformation Mechanisms

The manufacturing processes like rolling, forging, extrusion, drawing etc. completely rely on controlled plastic deformation for metals to realize the desired size and shape of an engineering component. The plastic deformation achieved by applying stress greater than yield strength of the metal under process. In general, deformation occurs by two mechanisms, namely slip, twinning or in combination of these two.

1.15.1 Slip

The movement of one layer of atoms over another layer needs shear stress above a critical stress level to cause a slip. Slip takes place through multiple inter-atomic distances; therefore, the arrangement of atoms in deformed and un-deformed regions remains same (Fig. 1.17). The slip results in step. The width of the step developed depends on the extent of slip / movement of sliding layers of atoms.

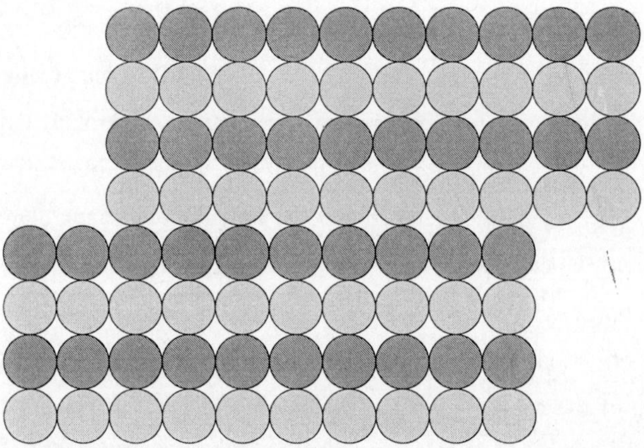

Figure 1.17 Schematic diagram of deformation due to slip

1.15.2 Twinning

The twinning mechanism of plastic deformation of metals involves the movement of one layer of atoms over another layer by a fraction of inter-atomic distance; therefore, the twinned zone shows a different orientation of atoms than the untwined region. Slip, on the other as mentioned earlier, causes movement of layer of atoms on the slip plane by one or multiple of inter-atomic distance therefore, both unreformed zone and zone deformed by slip both exhibit the same orientation of atomic arrangement (Fig. 1.18). The deformation of metals under low temperature and high strain rate conditions mainly occurs by twining. In reality, the deformation of metals usually occurs by both slip and twinning mechanisms.

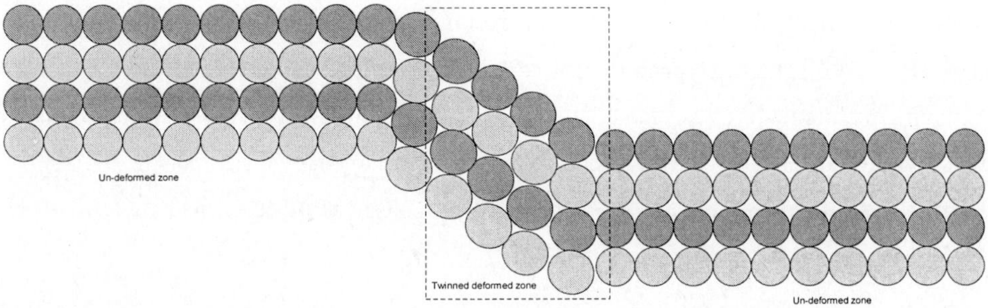

Figure 1.18 Schematic diagram of deformation mechanism by twinning

1.16 Critically resolved shear stress

The plastic deformation of crystalline metals by atomic level mechanisms namely slip and twinning takes place by sharing action on certain crystallographic planes in few select directions when shear stress component (of externally applied force exceeds) a critical shear stress level. The most closed packed atomic planes and in certain preferred directions called slip and direction respectively show a high slip/twining tendency to cause the plastic deformation. The relationship crystallographic planes with axis of a unit cell is represented Miller Indices. The reciprocal of the intersection of any such crystallographic plane with axis used to identify the specific planes in terms of miller indices (Fig. 1.19).

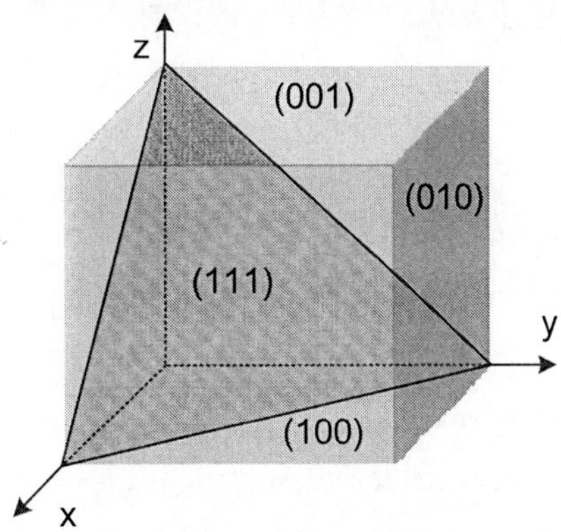

Figure 1.19 Schematic diagram showing a variety of planes in unit cell

Let's first consider stress condition leading to plastic deformation in a metal of single crystal to understand the deformation behaviour of poly-crystalline metal. Assuming that the single crystal has many dislocations. On application of tensile load "P" slip occurs on a slip plane, and a step of one / multiple inter-atomic width produced. Likewise, slip on other slip planes will continue at higher stress levels due to less favorably oriented slip planes/directions.

If the slip plane is inclined at "Θ" degree, then to establish the shear stress component that cause slip externally applied tensile stress is resolved in two components a) along the shear plane (Fs: PcosΘ) and b) perpendicular to the shear plane (Fn: PsinΘ) as shown in Fig. 1.20. Assuming the normal cross-sectional area of the components is A, then the shear area of the plane inclined at "Θ" degree with respect to axis of external load will be A/sinΘ.

This the shear stress on shear plane, Fs : (PcosΘ)/ (A/sinΘ) = (P/A)cosΘ.sinΘ

This the stress normal to the shear plane Fn: (PsinΘ)/ (A/sinΘ) = (P/A)$\sin^2\Theta$

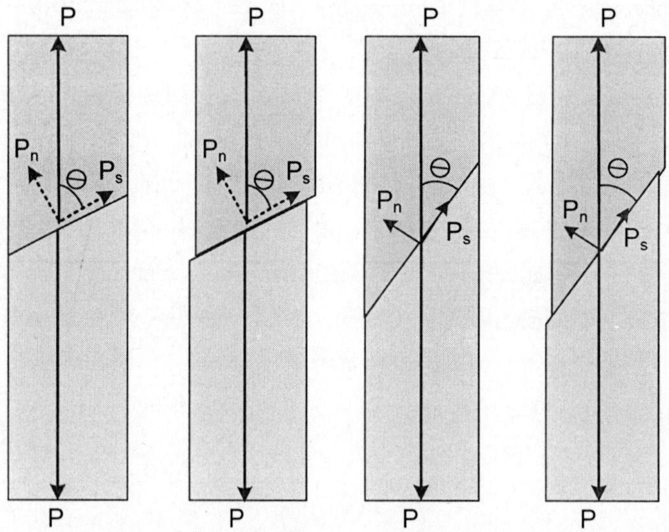

a) before b) after c) before d) after

Figure 1.20 Schematic diagram showing resolved components of forces on two different slip planes (before and after slip) in case of a-b) low angle of slip plane with respect to normal to external load before and c-d) high angle of slip plane with respect to normal to external load on a single crystal subjected to tensile load

The above equation of shear stress on the shear plane suggests that the maximum shear stress components due to external tensile loading occurs when the shear plane is inclined at 45°. In addition to the slip plane angle (with respect to the external loading), the most preferred slip direction (direction of most closed packed atoms) offering minimum resistance to the slip is another aspect affecting the minimum shear stress level desired to cause the slip (Fig.1.21a). The minimum shear stress level desired to cause the slip, called critically resolved shear stress, needs consideration of the slip direction with respect to the shear stress component (Fs) of the external load. Shear stress in preferred shear direction Fsr: Fs cosλ: (PcosΘ).cosλ / (A/sinΘ) = (P/A)cosΘ.sinΘ.cosλ

Analysis of shear deformation of single crystal by slip suggests the following two aspects:

a) External stress to cause deformation by slip through shearing action is minimum when the shear plane and corresponding shear direction exist at 45°

b) Stress requirement increases to cause plastic deformation by slip if shear plane and shear direction are located at other angles (higher or lower than 45°)

The polycrystalline metals having hundreds to thousands of crystals (even of the same crystal structure) with different orientation of atomic arrangements in each crystal (grain) so the other slip planes and slip directions (Fig. 1.21b). The most preferred plane to cause slip in each crystal will be located at different angles with respect to the externally applied load. Therefore, polycrystalline metals (under normal ambient conditions) need higher stress to trigger the slip and cause the plastic deformation than the single crystal. The critical resolved shear stress to initiate the slip in polycrystalline metal is higher than a single crystal. Further, deformation by slip within a crystal/grain is limited up to grain boundary. Continuous deformation across the section needs that slip triggers and progresses in adjacent grains. Hence, polycrystalline metals offer greater resistance to the plastic deformation and higher yield strength.

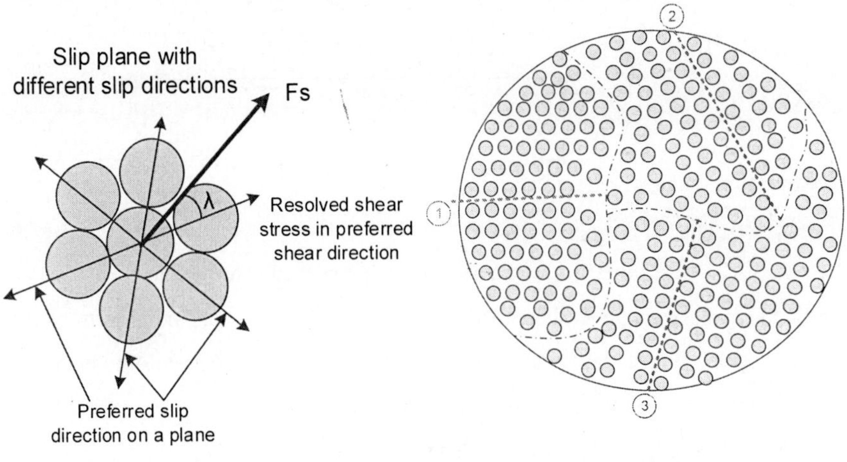

a) b)

Figure 1.21 Schematic diagram showing a) slip systems and resolved components of shear force for HCP metal and b) polycrystalline metals having different orientation of atomic arrangements in each grain causing different preferred angle of slip planes

The crystal structure affects the number of slip system (product of slip plane and slip direction) in a unit cell. The presence of fewer number of slip systems (8) in BCC and HCP metals leads to high critical resolved shear stress than FCC metals (Table 1.1). HCP metal like Mg having very few slip planes and directions, however the deformation (under the influence of external stress) by twinning assists in getting more slip system in a favorable position to facilitate the slip, which in turn helps to realize plasticity almost similar to that of FCC metals and higher than BCC metals. BCC metals having fewer slip planes and very few closed packed atomic direction resulting in very high critical resolved shear stress.

Table: 1.1 Critically resolved shear stress for different metals

Sr. No.	Metal	Crystal structure	Critically resolved shear stress, MPa
1	Ag	F.C.C.	0.37
2	Cu	F.C.C.	0.49
3	Al	F.C.C.	0.78
4	Mg	H.C.P.	0.44
5	Co	H.C.P.	6.62
6	Ti	H.C.P.	13.10
7	Fe	B.C.C.	27.44
8	Nb	B.C.C.	33.37
9	Mo	B.C.C.	71.70

1.17 Dislocation Theory

The perfect crystals offer high resistance to deformation as per stress required to break to atomic bonds. Stress required for deforming a perfect crystal is termed as theoretical strength. However, the real crystal general comprises many crystalline defects in the form of vacancies, dislocations, grain boundaries, cracks etc. Therefore, slip of real crystalline metals occurs at much lower (100 to 1000 times) stress than the theoretical strength. This difference in actual and theoretical strength explained with the help of dislocation theory.

As mentioned earlier, crystallographic imperfection (in real metals) dislocations are two types i.e. edge and screw dislocation. These dislocations under the influence of external stress moves at favorable orientated slip plane. If moving dislocations comes across an obstacle / barrier then slip starts on other favorably oriented parallel slip planes. The movement of edge dislocations occurs perpendicular to slip plane while screw dislocations move parallel to the slip plane. Edge and screw dislocations move systematically toward the end or grain boundary and produce a step (Fig.1.22)

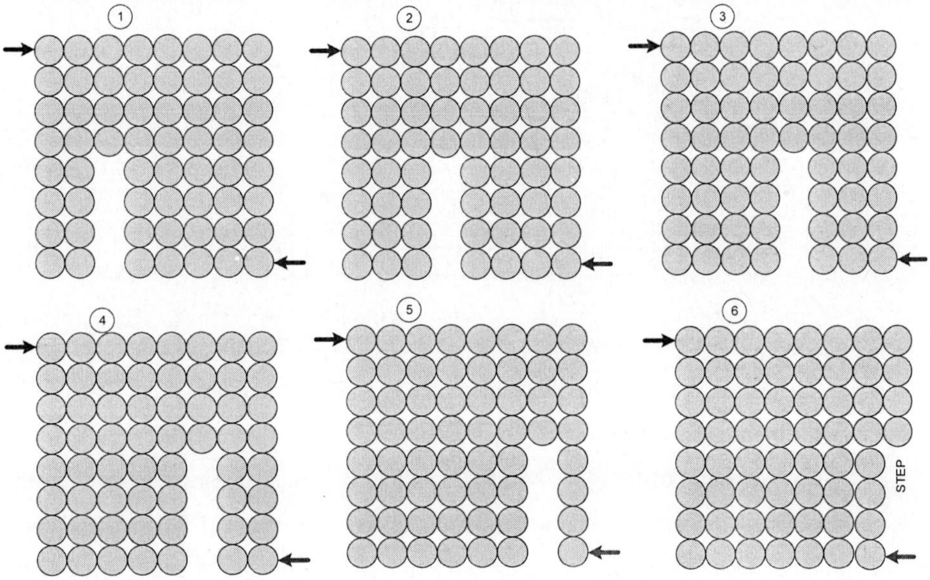

Figure 1.22 Schematic diagram showing mechanism of deformation by slip due to dislocation movement

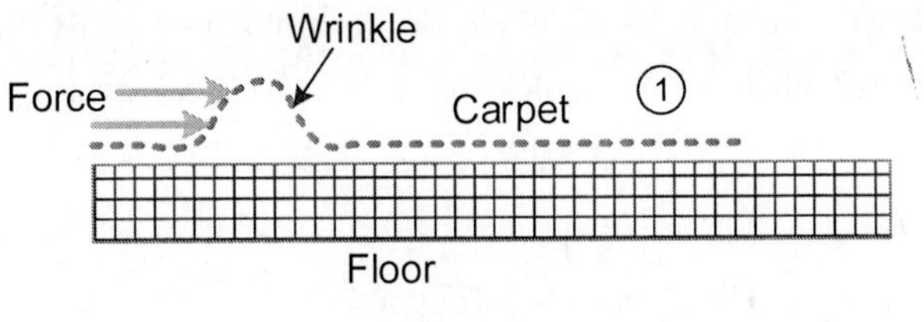

a) Wrinkle at end moved by little force

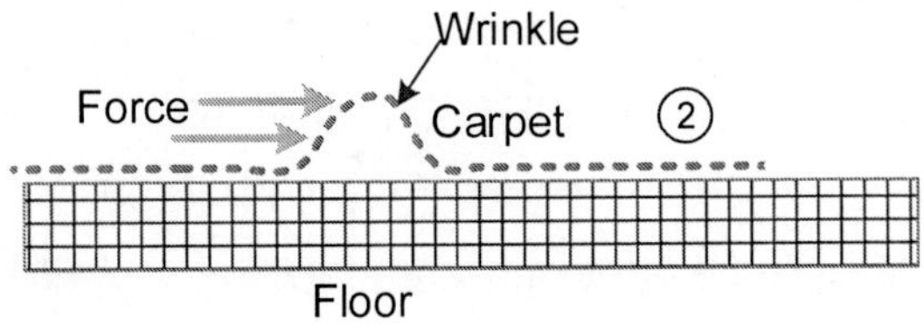

b) Wrinkle moving toward right

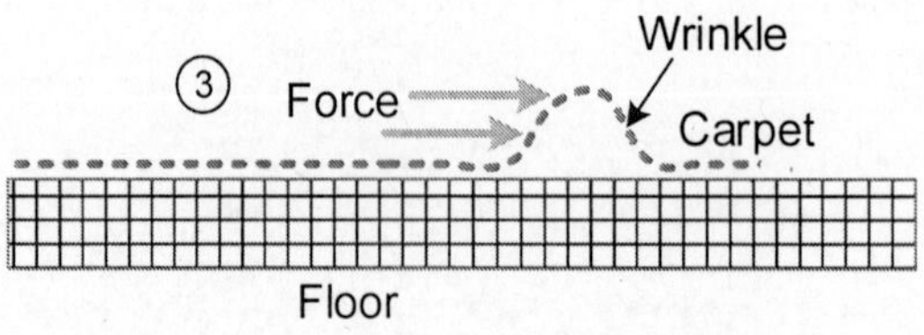

c) Wrinkle movement continues

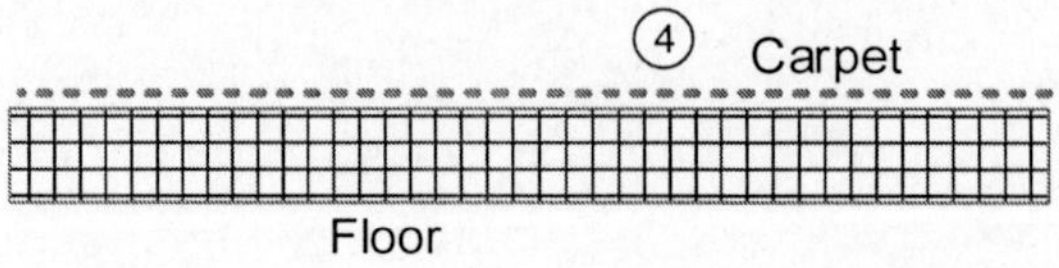

d) Wrinkle reaches at end on right and moves carpet accordingly

Figure 1.23 Schematic diagram showing carpet analogy for dislocation movement at very low stress than the theoretical strength

The movement of dislocations occurs at much lower shear stresses than the theoretical stresses obtained on the basis on atomic bond strength. The movement of dislocation is similar to the motion of a wrinkle in the carpet from one location to another or end. As the wrinkle reaches to the end of carpet, the carpet moves by some distance. The force / stress required to move wrinkle is much lower than that in needed pulling entire carpet by same distance (Fig 1.23). The wrinkle in this example is representing a dislocation.

Edge dislocations are considered as an extra incomplete plane of atoms sandwiched between two complete planes of atom. Presence of an edge dislocation develops a stress field in its vicinity. The side in which extra incomplete plane of atoms is present, experiences compressive stresses while another side from where plane of atoms is missing, experiences tensile stresses. Thus, based on location of the dislocations with respect to the slip plane these can be classified as positive or negative dislocation (Fig. 1.24). As these dislocations move, stress field is shifted with them. Any kind of interference of this stress field of moving dislocation with "other stress" fields resists the movement of dislocations. These "other stress" fields may be due to foreign solute atoms or other dislocations present close to the moving dislocations. Increase in the extent of interfering effect of these "other stresses", increases the external stresses required to cause the slip. This concept in used to strengthen material by work/strain hardening mechanism by producing large number of dislocations via severe plastic deformation of stock materials under consideration.

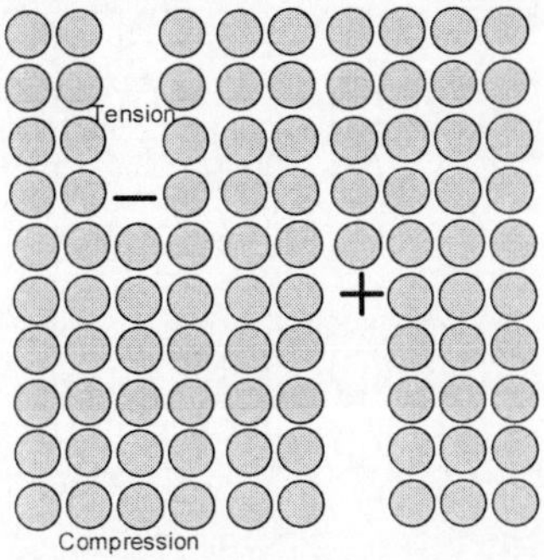

Figure 1.24 Schematic diagram showing characteristics of dislocations and their effect in lattice strain

Edges of dislocations can move up or down by diffusion of atoms toward or away from the edge of the incomplete plane of atoms (Fig. 1.25). This phenomenon called ***climb up or down***. When edge of an incomplete plane of atoms moves up due to loss atom it is called climb up. Similarly, an addition of atoms to an edge of incomplete plane of atoms moves it down. This is phenomenon is termed as climb down. Climb up/down is based on the diffusion and occurs rapidly at high temperature and changes the slip planes easily, to facilitates the movement of dislocations on other plane when some barriers obstruct them at the same stress level. Climb up/down plays a significant role in high temperature deformation of metal at constant load with time called ***creep***. Therefore, dislocation movement at high temperature becomes easier than room temperature as climb up /down help to bypass the obstacles like precipitates, and other hardening micro-constituents.

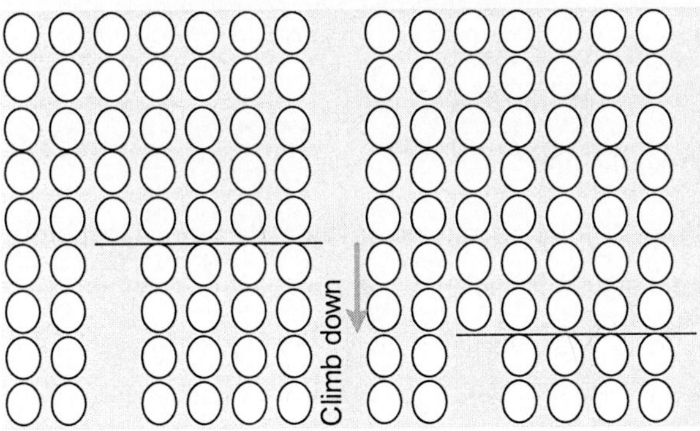

a)

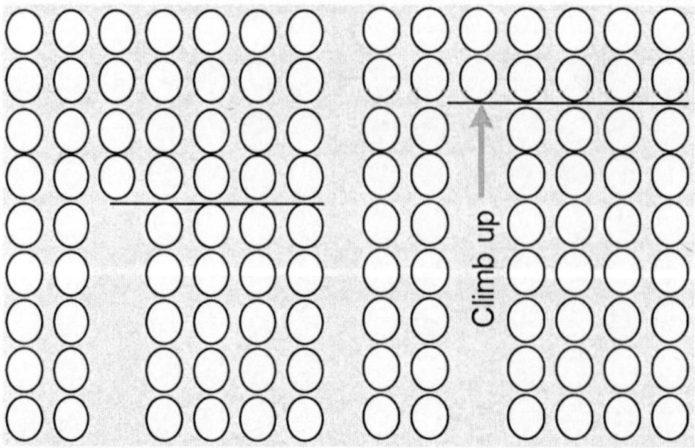

b)

Figure 1.25 Schematic diagram showing the effect of a) climb down and (b) climb up

1.18 Metal Strengthening Mechanism

The dislocation theory suggests that all the factors, which can act, as obstacles for the dislocation movement will in turn increase the resistance to the plastic deformation and conversely increase the yield strength and hardness. Presence of fine grains leading to many grain boundaries and increased interfering effect of stress field (due to lattice distortion) due to presence of many other dislocations, precipitates, alloying elements as solute, hardening constituents in the proximity / vicinity of the moving dislocations all increase the resistance to dislocation movement. Therefore, all these can be used (singly or in combination) to strengthen the metals using common approaches (based on above fundamentals) namely grain refinement, work hardening, precipitation hardening, solid solution strengthening and dispersion hardening. Additionally, transformation of soft phases to hard phases (in few metals like alloys of Fe, Cu, Ti, Co etc.) using a controlled thermal cycle or plastic deformation are another type of approaches called transformation hardening.

1.18.1 Grain refinement Strengthening

A large grains in a metal allows dislocation movement along the particular slip plane over a long distance (in terms of inter-atomic distance) maximum equal to the grain size provided, there is no obstacle for the movement. The slip cannot continue on the same slip plane in adjoining grains due to different orientations of atoms even in case of the same crystal structure metals. Therefore, dislocations continue to get plied up near the grain boundary, which eventually triggers the slip in neighbouring grain to facilitate the deformation. On the other hand, the extent of movement of dislocations (in terms of inter-atomic distance) in fine grains is limited maximum up to grain boundary. In case of fine grain metals, the piling up of the dislocations at the grain boundaries is very limited as compared to that of the fine grain metals, which in turn reduces the plastic deformation tendency in fine grain metals (Fig. 1.26). Therefore, refinement of grains in a metal results in increased grain boundaries, which in turn causes increased obstacles for dislocation movement. Thus, the grain refinement increases the yield strength. The yield strength of a metal is inversely related with square average grain size. The Hall Patch equation showing the relationship between the yield strength and grain size is shown below.

$$\sigma_y = \sigma_i + k / d^{1/2}$$

Where σ_y is the yield strength of polycrystalline material, σ_i is the yield strength of a single crystal of infinite size (lattice friction), k is constant and d is average grain diameter. Grain refinement increases the yield strength & fatigue strength and hardness without loss of ductility.

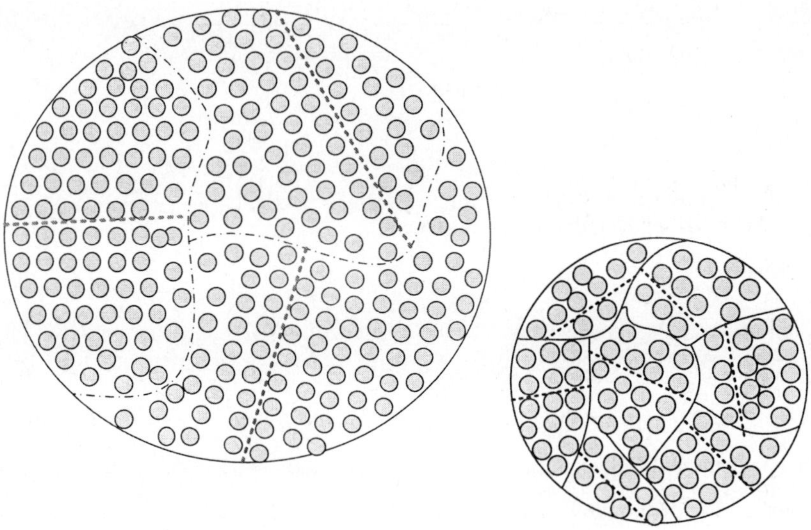

a) b)

Figure 1.26 Schematic diagram showing the effect of grain size on slip of a) coarse grain (b) fine grain

In general, a decrease in grain size (at micro-level) increases the yield strength of metal. However, reduction in grain size below 50- 10 nm reduces the yield strength and a reversal of Hall-Patch relationship is observed (Fig. 1.27a). Many hypotheses have been proposed to explain reversal of Hall-Patch relationship namely Coble creep tendency (creep due to diffusion of grain boundaries), grain boundary sliding and absorption of dislocation by grain boundaries. Further, the extent of increase in yield with a reduction in grain size is somewhat lesser than other approaches. An attractive aspect of strengthening by grain refinement is that yield strength is enhanced without compromising ductility and toughness, however, creep resistance/ strength is adversely affected due to increased grain boundary sliding tendency (Fig. 1.27).

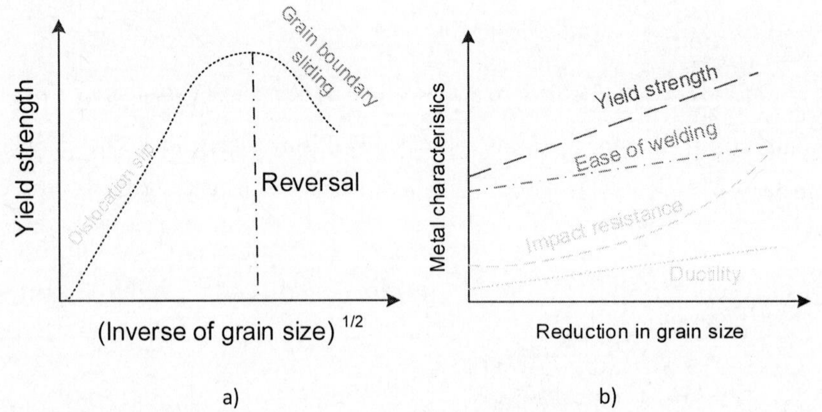

Figure 1.27 Schematic diagram showing the effect of grain size on a) yield strength as per Hall Patch relation and (b) various characteristics of metals

1.18.2 Strain Hardening

A perfect crystal offers maximum resistance to the dislocation movement. However, the real metals usually have few dislocations, which weaken the metal by facilitating easy dislocation movement (as per dislocation theory) to cause plastic deformation. The presence of dislocations in metals develops the strain fields in the proximity of dislocation. The strain field around the dislocations interferes with the moving dislocations. Like same type (positive / negative) of dislocations repel while opposite types of dislocation attract each other. Increased number of dislocations increases the chances of such interferences with the moving dislocations. Therefore, a controlled plastic deformation using a suitable approach like rolling, forming etc. produces dislocation (Fig. 1.28). Increased plastic deformation of metals increases the dislocation density, which in turn increases the resistance to the dislocation movement so yield strength of the metals. An increase in the yield strength of a metal is directly proportional to the square root of number of dislocations.

$$\tau_y = \tau_0 + a\,\rho^{1/2}$$

where τ_y is shear stress required to cause the deformation with ρ dislocation density, τ_0 is shear stress required to cause the deformation with zero dislocation density (annealed condition) and a is constant.

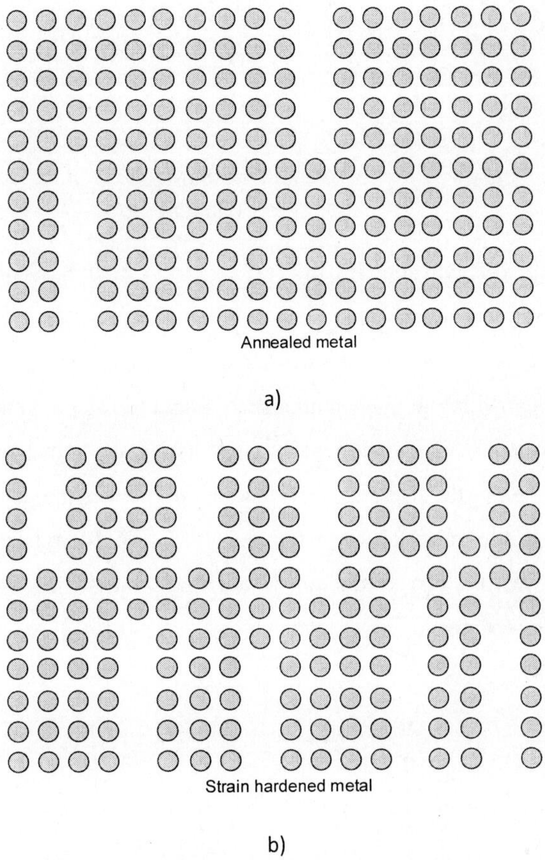

Figure 1.28 Schematic diagram showing dislocations in a) annealed metal and b) plastically deformed to cause strain hardening

Indirectly, the annealing of metals (high temperature exposure for a long time) reduces the dislocation by annihilation, climb up/down which in turn leads to the softening of metals (Fig.1.27). Increase in strength of metals due to work hardening significantly depends on the metal system. A metal with low stack fault energy (Co, Ti, stainless steel, super alloys, steels) responds in a bigger way than those of high stacking fault energy (Mg, Al). New dislocations act as a barrier for the movement of already existing dislocations (Fig. 1.29).

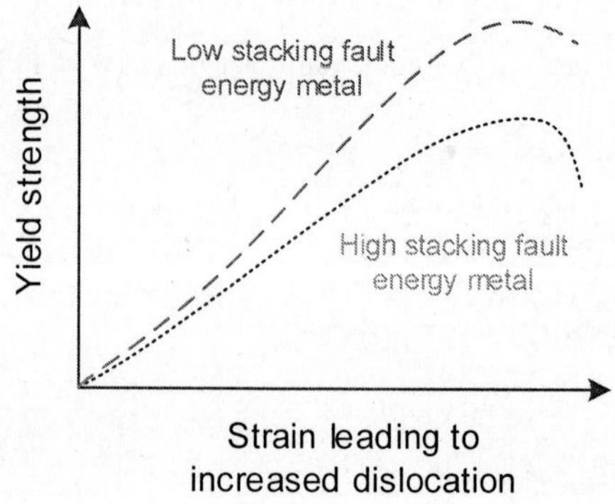

Figure 1.29 Schematic diagram showing the varying effect strain hardening on metals having different stacking fault energies

Strengthening of the metal by work hardening used in deformation based manufacturing processes like forging, rolling to make components (hand tools and connecting rods) of high strength to weight ratio which in turn helps in reducing the metal cost and dimensions. However, strengthening of the metal by work hardening reduces the ductility, toughness and increases ductile to brittle transition temperature (DBTT). DBTT indicates a narrow band of temperature below which a sharp reduction in toughness takes place and is crucial for the metallic components expected to work under sub-zero conditions (1.30).

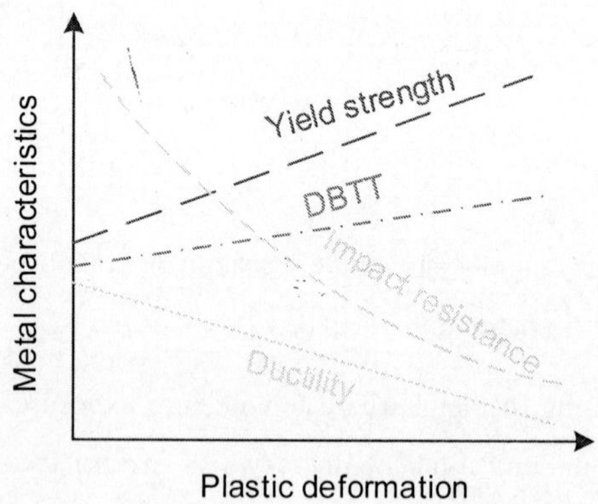

Figure 1.30 Schematic diagram showing the effect of plastic deformation/strain hardening on various mechanical characteristics

1.18.3 Solid solution strengthening

The pure metals are soft, ductile and low strength and therefore not found very useful to manufacture products for engineering and other applications. For example, pure aluminium offers yield strength around 50 MPa while many aluminium alloys designed suitable offer strength 300-350 MPa. This is realised by controlled alloying of pure metal to achieve high strength through solid solution strengthening and precipitation hardening. The presence of an element (metal or non-metal) in a metal matrix causes distortion of lattice due to usually misfit atomic size (due to atomic size difference) of alloying element (as solute) added in the metal matrix to form a solid solution (Fig. 1.31). Alloying elements can form substitutional solid solution or interstitial solid solution. The substitutional solid solution is formed when atoms of an alloying element displaces the atoms of the metal matrix while in case of interstitial solid solution, atoms of alloying element occupy the interstitial spaces between the atoms of metal matrix. Interstitial solid solution generally causes more lattice distortion than substitutional solid solution. Accordingly, the mechanical properties of metal matrix are affected by the alloying elements. The extent misfit in atomic size of alloying elements and matrix determines the degree of lattice distortion. The lattice distortion in turn develops a strain field around the alloying elements, which interferes with moving dislocations and increases the resistance to the movement of dislocations and so the yield strength enhanced. An increase in misfit increases the lattice distortion, which in turn increases the extent of hardening and strengthening of metal due to alloying. Therefore, different alloying element (even with same wt. %) offers a varying increase in strength of the metal matrix (Fig. 1.32). In general, increasing % of alloying element increases the strength of the metal matrix. Therefore, effect of allying elements on yield strength of metal depends on a) extent of atomic misfit of alloying element and the metal matrix, b) type of solution formed and c) amount (wt.%) of alloying element.

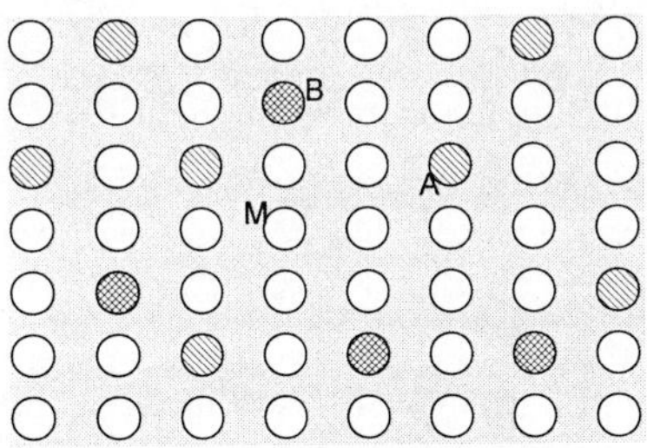

Metal matrix, M, Alloying elements A & B

a)

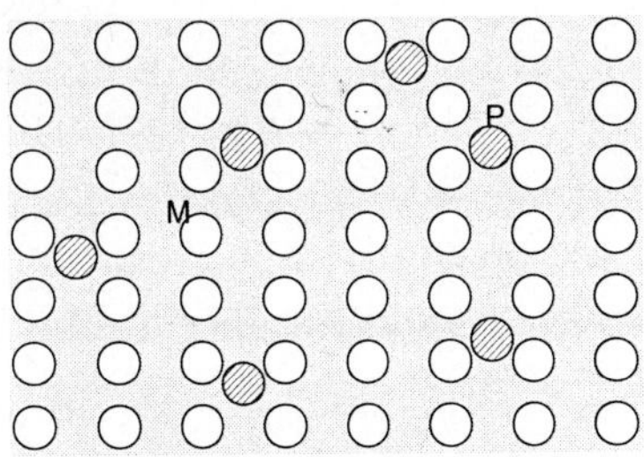

Metal matrix, M, Alloying element P

b)

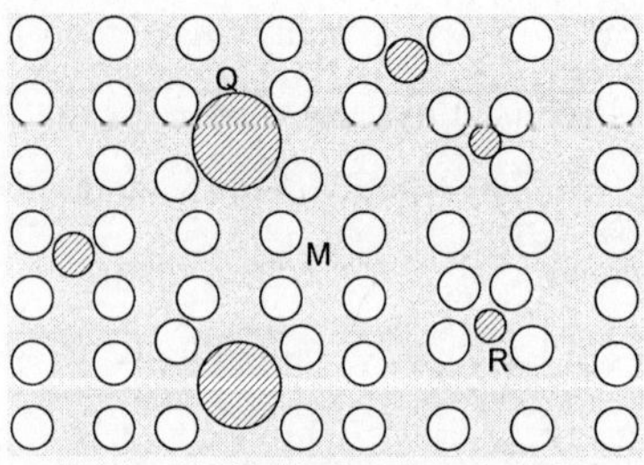

c)

Figure 1.31 Schematic diagram showing solid solution a) substitutional type, b) interstitial type and c) interstitial solid solution showing misfit

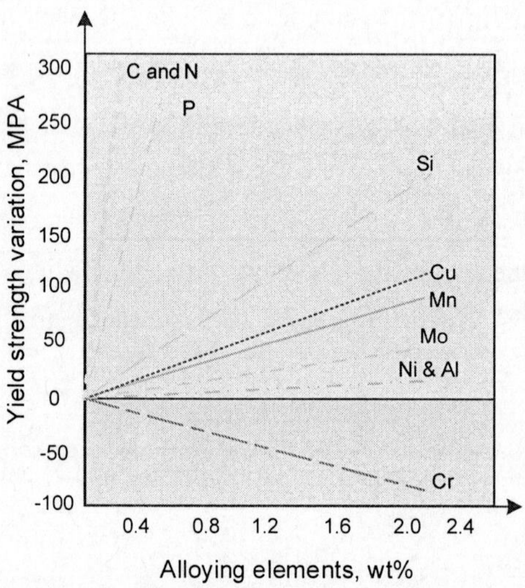

Figure 1.32 Schematic diagram showing varying effect of alloying element on yield strength of Fe

1.18.4 Precipitation hardening

The precipitation hardening is based on the ability of fine and well-distributed coherent precipitates to effectively act as obstacles for the movement of dislocations. In metal matrix, the presence of precipitates strain field around the precipitation causes the lattice distortion, which in turn helps to resist the movement of dislocations and so enhances the strength of the metal. The formation of precipitation depends on the variation in solid-state solubility of alloying elements in the metal matrix as a function of temperature. The solubility of alloying elements in the metal matrix increases with the temperature while it decreases with the reduction in temperature (Fig. 1.33).

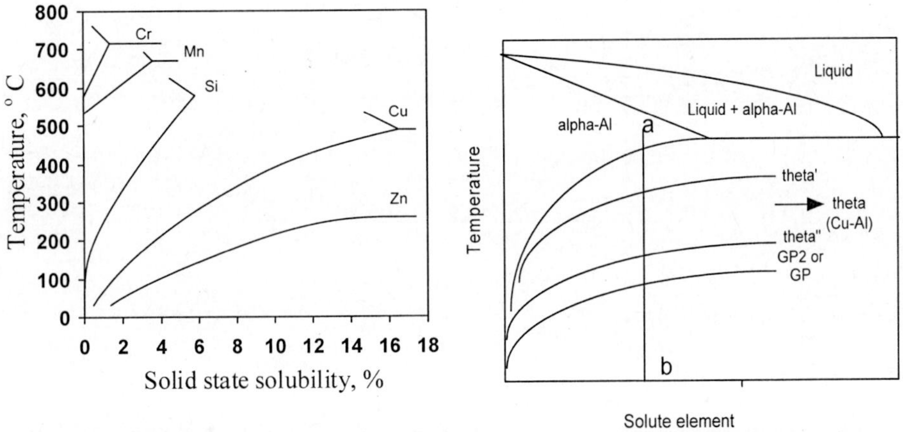

a) b)

Figure 1.33 Solid solubility of alloying elements in aluminium as a function of temperature

Precipitation hardening of suitable alloy showing above behaviour with regard to solubility, performed in three steps namely solutionizing, quenching and aging. The solutionizing involves heating of the alloy to high enough temperature (above the solvus line in in phase diagram and below the eutectic/melting temperature) followed by holding for some time (30 min to 4 hrs) as per metal system and section thickness. The quenching (rapid cooling) of the alloy from the solutionizing temperature results in the formation of super-saturated solid solution of the alloy by diffuseness process. Aging of quenched alloys performed at room temperature is called as natural aging, and aging done at high temperature (150-1000 °C) as per metal system is known as artificial aging. During the aging step only excess solute (as per aging temperature for a metal system) from the matrix is rejected and eventually interaction of solute with matrix metal produces fine and well distributed hardening precipitates (Fig. 1.33).

The precipitation can occur both within grains and at the grain boundaries. The kinetic of precipitation accelerated at high temperature during artificial aging, and peak hardness/strength achieved faster than natural aging. Aging condition (temperature and time combination) corresponding to the peak hardness is called peak aging condition. Over and under exposure (in terms of time) of alloy during aging results in lower hardness/strength than the peak aging and accordingly it is called under-aged or over-aged alloy (Fig.1.34).

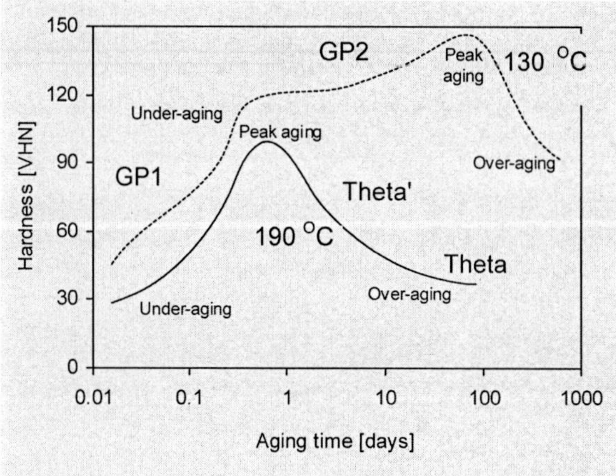

Figure 1.34 Effect of aging time and temperature on hardness of PH aluminium alloys

For example, Al-Mg, Al-Cu alloys form Al_3Mg_2 and $CuAl_2$ hardening precipitates. These appear as fine well-distributed precipitates in aluminium matrix, which act as a barrier for the movement of dislocations and increases the strength and hardness. Fine precipitates are more effective than the coarse one in increasing strength. The movement of dislocation in precipitation-hardened alloys occurs by two mechanisms a) cut through precipitates and b) by passing precipitates of depending upon their size. Fine precipitates primarily determine the strength of an alloy. Cut through mechanism dominates with precipitates having size more than average critical radius of precipitates otherwise dislocation bypass mechanism dictates the strength in case of fine precipitates (Fig.1.35).

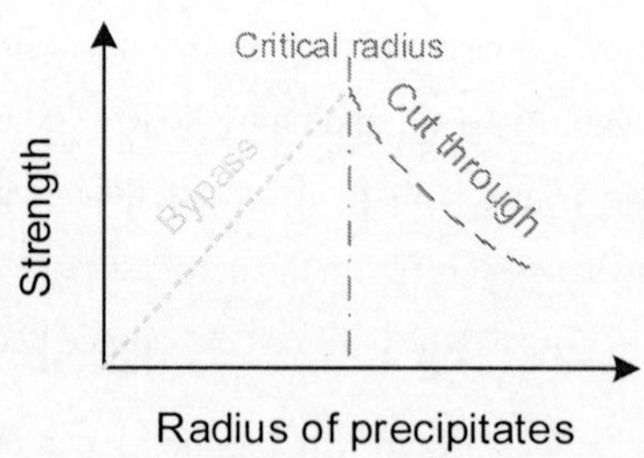

Figure 1.35 Schematic diagram showing the effect of grain size on yield strength and corresponding mechanisms of dislocation movement across the precipitates

The extent of increase in strength and hardness after precipitation hardening depends on the kind of precipitates formed. Precipitates can be coherent or non-coherent (Fig.1.136). Coherent precipitates produces large stress field (lattice stain around precipitates) which effectively hinders the movement of dislocations to increase the hardness and yield strength. On contrary, non-coherent precipitates do not contribute much on strengthening of alloys.

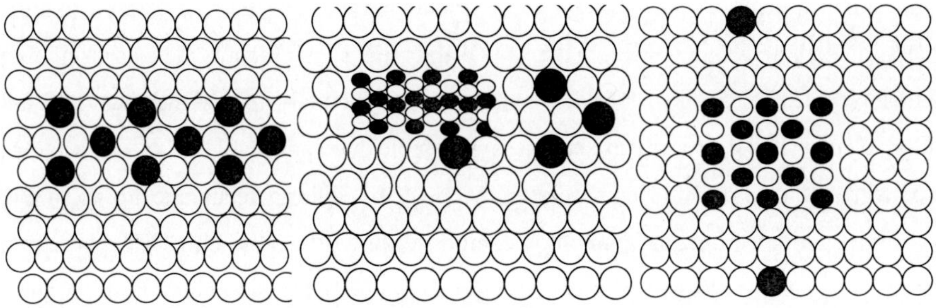

a) b) c)

Figure 1.36 Schematic diagrams of a) coherent, b-c) non-coherent precipitates

Five sequential structural transitions occur during the artificial aging of Al-Cu alloys:

$$SS \longrightarrow GP \longrightarrow \theta'' \longrightarrow \theta' \longrightarrow \theta\ (CuAl_2)$$

Where θ is an equilibrium phase. The GP zones (Guinier-Preston, sometimes called GP1), the θ'' phase (sometimes called GP2), and the θ' phase are meta-stable phases. The solvus curves of these meta-stable phases shows the highest temperatures up to which these phases are stable and above that they get dissolve in the matrix.

The GP zones are coherent with the crystal lattice of solid solution. These GP zones consist of disks a few atoms thick (4-6A) and about 80-100A in diameter. The 'θ'phase is also coherent with the crystal lattice. The severe lattice strains associated with the coherent precipitate make the movement of dislocations more difficult. Depending upon the alloy system, the following reactions can take place during the aging process.

- Al-Cu : SS ~ GP ~ S' (Al_2CuMg) ~ S (Al_2CuMg)
- Al-Mg-Si: SS ~ GP ~ ß ' (Mg_2Si) ~ ß (Mg_2Si)
- Al-Zn-Mg: SS~ GP ~ γ' (Zn_2Mg) ~ γ (Zn_2Mg)

1.18.5 Dispersion hardening

Metals strengthening mechanisms namely grain refinement, precipitation hardening and strain hardening are not effective to increase the strength and hardness for high temperature applications primarily due to structural instability in the form of grain coarsening, recovery and recrystallization, tempering, over-tempering/aging, reversion (dissolution and coarsening of precipitates) experienced of metals at elevated temperatures. These microstructural changes at elevated adversely affect the mechanical properties, including strength and hardness. The dispersion hardening is an effective approach to strengthen the metals by reinforcing hard and stable micro-constituents in the soft and tough metal matrix. The hard and stable micro-constituents do not interact much with matrix metal and retains their identity. The hard and fine micro-constituents like carbide, borides and oxides effectively pin down the moving dislocation to enhance the strength and hardness. Insoluble, hard and stable micro-constituents are mixed with metal matrix and then processed by powder metallurgy technique. For example, silicon carbide reinforced in aluminium alloy composite shows the presence of SiC particle in the matrix, is an example of dispersion hardening of aluminium. Sintered aluminium products (SAP) are another example of dispersion hardening where alumina is reinforced in aluminium matrix. Similarly, TiC reinforcement in copper matrix increases life of spot welding electrodes appreciably without much loss of desired thermal and electrical conductivity.

1.19 Microstructure

Microstructure of a metal reveals the phase structure and grain structure. Phase structure shows the type of phases, relative amount of various phases present in an alloy and distribution of phases. A phase represents all micro-constituents in a given metal having an identical composition, crystal structure and state. Any difference in the above three parameters results in formation of new phase. The new phase may have its own crystal structure, state and composition. Grain structure reveals the size and shape of grains and their distribution. Grains may be of different shapes such as lamellae, needle, spherical (compact or irregular), flake, cuboid, polyhedral, star and Chinese etc. (Fig.1.37). Each shape has a different impact on mechanical properties like fatigue, fracture toughness, tensile strength, and ductility.

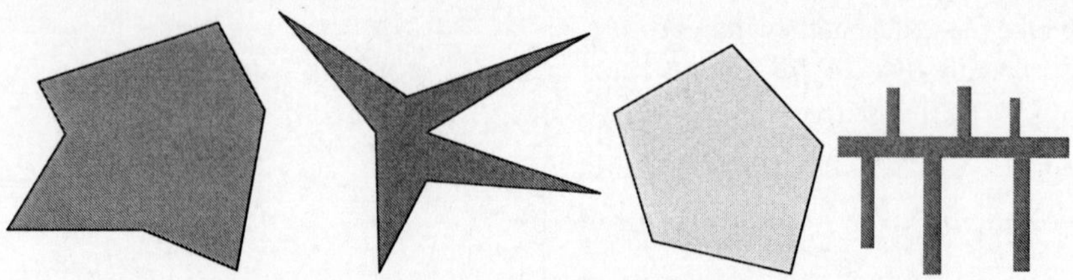

a) b) c) d)

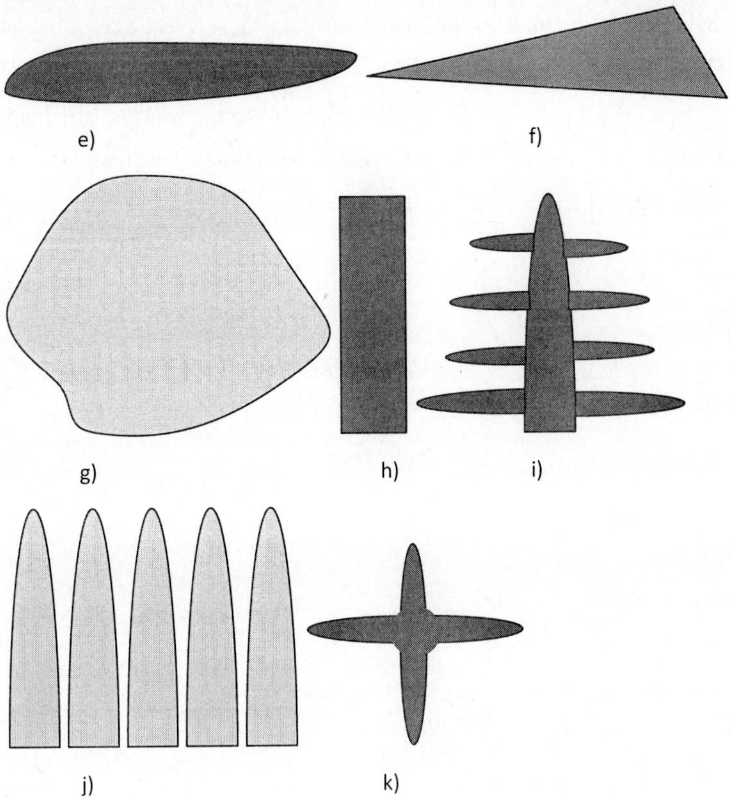

Figure 1.37 Schematic diagram showing typical grain morphologies / shapes a) irregular, b) star, c) polyhedral, d) Chinese script, e) flake, f) needle, g) circular family, h) planar, i) dendrite, j) cellular and k) equiaxed

1.20 Structure-Property-Correlation

Structure in physical metallurgy is generally termed as microstructure. The microstructure and mechanical property correlation has been well established. Therefore, attempts are made first to control/modify the microstructure to obtain desired mechanical properties. Microstructure can be modified by changing either its phase structure or grain structure. The modification of microstructure can be obtained using some of the following methods such as changing a) the composition, b) processing method (casting, forming, and machining) and c) heat treatment conditions as per requirements.

UNIT SUMMARY

The development of materials, common materials of commercial importance for the current time, fundamentals of crystal structure and its effect on properties of materials like metal, ceramic, and polymers have been elaborated. Additionally, the atomic level plastic deformation mechanism like slip and twining have been presented. The fundamental of metal strengthening mechanisms namely grain refinement, solid solution strengthening, strain hardening, precipitation hardening have also been explained using suitable schematics. The importance of microstructure and its effect on properties have been briefly covered.

EXERCISE

Multiple Choice Questions

Questions for self-assessment

1. Long molecular chains are found in
   a. Metals
   b. Ceramic
   c. Composite
   d. Polymers
2. Metal tends to deform easily if the number of slip systems is
   a. 12
   b. 8
   c. 4
   d. 0
3. Effective number of atoms are maximum with metals having
   a. HCP
   b. CCP
   c. BCC
   d. FCC
4. Free electrons in metals increases
   a. Electrical resistivity
   b. Thermal conductivity
   c. Density
   d. Strength
5. The atomic packing factor is maximum for
   a. Al
   b. Zn
   c. Fe

   d. All metals

6. Density of a metal will be maximum of Ceramic are known to possess high
   a. Density
   b. Hardness
   c. Toughness
   d. Ductility
7. With grain refinement, a metal shows
   a. Increase in yield strength only
   b. Decrease in yield strength only
   c. No change in yield strength
   d. Yield strength first increases then decreases below the critical size
8. Increasing of crystallinity in polymers increases
   a. Strength
   b. Density
   c. Melting point
   d. All these
9. The maximum yield strength is observed when there is
   a. No crystalline defect
   b. Few dislocations
   c. Few grain boundaries
   d. No precipitates
10. Strengthening by work hardening of carbon steel results in
    a. Decrease of strength
    b. Increase of toughness
    c. Decrease of DBTT
    d. Increase of ductility

Answers of Multiple Choice Questions

Key for MCQ: 1 d, 2 a, 3 a, 4 b, 5 b, 6 b, 7 a, 8 a, 9 a, 10 d

Short and Long Answer Type Questions

1. How did the importance of materials change historically over a long time horizon?
2. Enlist the materials of commercial importance
3. What is unit cell?
4. Distinguish the crystalline, amorphous and allotropic materials

5. Define effective and coordination numbers related to a unit cell
6. What arc common crustal structures of metals?
7. How does the crystal structure of metals, polymers and ceramic materials differ from each other?
8. How does the crystal structure affect the mechanical and physical properties of materials?
9. Define the atomic packing factors and write its significance.
10. Explain the plastic deformation mechanism of metals by slip and twinning.
11. Describe dislocation theory of plastic deformation by slip.
12. What are metal strengthening mechanisms?
13. What are crystal defects?
14. Explain the various crystalline defects using suitable schematics.
15. How does the volume defects affect the strength of metals?
16. Describe the grain refinement and precipitation hardening metal strengthening mechanisms.

Numerical Problems

1. Determine the coordination and effective number of atoms of Al, Zn and austenitic stainless steel, alpha-ferrite.
2. Calculate the atomic packing factors of Cu, Fe, and Mg.

PRACTICAL

1. Study the effect of plastic deformation on the mechanical properties of metals.
2. Study the effect of natural and artificial aging during precipitation hardening of aluminium 7xxx or 2xxx alloy.

KNOW MORE

Explore the ancient and historical objects of different materials and try to find how and why those have been sustained for a long time. For example, about 10 centuries old structure of pillar in Qutub Minar, Delhi, is still unaffected by atmospheric conditions while Eiffel Tower just 135 years old has started corroding and needs refurbishment.

REFERENCES AND SUGGESTED READINGS

Suggested resources for further reading/learning

1. Callister, Materials science and engineering, Wiley, Publication (2014)
2. V Raghan, Materials science and engineering, Wiley, Publication (2015)

3. H W Pollack, Materials science and metallurgy, Prentice Hall, (1988)
4. D K Dwivedi, Fundamentals of Metal Joining, Springer nature, Singapore (2022)
5. D K Dwivedi, Surface Engineering, Springer Nature, Singapore (2018)
6. Vivek Pancholi, NPTEL Course "Materials science and engineering": https://onlinecourses.nptel.ac.in/noc22_mm05/preview
7. Ratna Kumar, NPTEL Course "Basics of Materials Engineering" https://onlinecourses.nptel.ac.in/noc20_me78/preview

Dynamic QR Code for the Unit 1

SCAN ME

Unit 2

Mechanical Properties of Materials

Tensile, Compression, Torsion, Toughness and Hardness

Unit Specific / Learning Objective

Objective this unit in to talk about following aspects

- To introduce significance of mechanical properties under tension, compression, torsion, impact loading
- To learning about the importance of various mechanical properties in manufacturing and design
- To introduce the underlying principle of different mechanical properties evaluation
- To learns about important properties which can be obtained from different mechanical tests
- To introduce the concept of macro and micro-hardness
- To understand the significance of various parameters like modulus of elasticity and rigidity, yielding, yields strength, ultimate strength, ductility, resilience, off-setting strength
- To introduce the concept of the generalized hooks law
- To learn about estimation of tensile using hardness of materials

Additionally, few fundamental questions for self-assessment based on fundamentals **are** included in this chapter in form of recall, application, comprehension, analysis and synthesis. There are further suggested readings and reference for deep learners and readers assistance.

Rationale

Individual components, parts, or complete system are designed, manufactured for usage by public or manufacturing capital good itself. The performance of a product largely depends on choice of a suitable material for a given application (service condition) followed by its design and manufacturing. Selection of material therefore must be based on properties of materials required. The properties of materials significantly determines the approach of manufacturing and design of the product / component. Consideration of mechanical properties of materials therefore becomes inevitable for effective design and manufacturing of a reliable product. The development of understanding and evaluation of the mechanical properties plays a key role for mechanical engineers. In this unit, therefore, fundamentals related to mechanical properties of materials like tensile, compression, torsion, impact loading and hardness have been presented.

Pre-Requisites

A course on Physics: Force, stress, strain (Class XII)

Learning outcomes

U2-O1: Ability to choose suitable material based on mechanical properties for manufacturing and design

U2-O2: Ability to choose mechanical properties that should be examined for given application

U2-O3: Ability to interpret and estimate tensile strength of materials based on hardness

U2-O4: Ability to calculate various importance properties of materials like engineering & true stress/strain, yield and ultimate strength, ductility, resilience, modulus of elasticity and rigidity etc.

U1-O5: Ability to predict the behaviour of materials under impact and pointed loading using toughness and hardness respectively

Unit-1 Outcomes	EXPECTED MAPPING WITH COURSE OUTCOMES (1- Weak Correlation; 2- Medium correlation; 3- Strong Correlation)					
	CO-1	CO-2	CO-3	CO-4	CO-5	CO-6
U2-O1	1	2	3			
U2-O2	1	2	3			
U2-O3	1	2	3			
U2-O4	1	2	3			
U2-O5	1	2	3			

Course Objective

1. Student will be able to identify crystal structures for various materials and one could understand the defects in such structures.
2. Understand how to tailor material properties of ferrous and non-ferrous alloys?
3. How to quantify mechanical integrity and failure in materials?

2.1 Introduction

The mechanical property measurement is extremely important for the commercial exploitation of the materials and its products with full confidence and reliability. Establishing mechanical properties like tensile, compressive, torsion, toughness and hardness properties using a suitable test method is found to be very useful for production and design engineers. Based on the Mechanical properties of a material, the production engineers in consultation with the design engineers take a call on the appropriate manufacturing process, which can be used for making goods effectively, economically, and eco-friendly. On the other hand, the design engineers use suitable Mechanical properties like tensile strength, toughness, ductility, fatigue strength, creep resistance for designing the product suitable for given service conditions with regard to the load and environment conditions. Therefore, it is very important to characterize the mechanical properties of the materials.

Mechanical Properties

The mechanical properties of Material show its response to the external load conditions. The materials used for manufacturing products for human beings are subjected to various types of loads such as tensile, compression, shearing, bending, torsion, and impact or the combined one. The externally applied load may remain constant or fluctuate as a function of time and accordingly these loads are called static and dynamic loading. In the following section the objective and methodology of different mechanical tests namely tensile, compression, torsion toughness, and hardness test will be elaborated.

2.2 Tensile test

The tensile properties of the material are one of the most commonly used mechanical properties for designing components. Tensile properties of material show behavior of the material under externally applied tensile load conditions. The tensile test helps to establish the yield strength, ultimate strength, ductility in terms of percentage elongation and percentage reduction in area, and modulus of elasticity. The tensile test also helps to estimate the toughness of the material and its ability to undergo plastic deformation. Accordingly, the materials can be grouped as perfectly elastic, perfectly plastic, and elastic-plastic materials as shown in Fig. 2.1 below:

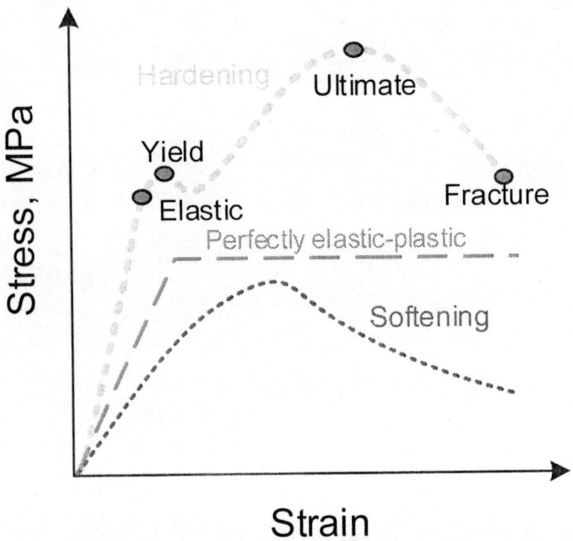

Figure 2.1: Schematic diagram showing stress vs. strain relation for different type of materials

A typical tensile test helps to establish the relationship between the tensile stress applied and the tensile strain experienced by the material during the tensile loading. The methodology of the stress can calculations helps to develop the two types of stress vs. strain diagram namely engineering stress versus engineering strain and true stress versus true strain. The engineering stress versus engineering strain diagram is developed using the initial stage sample dimensions namely the cross-section area of the tensile sample and the gauge length for calculating the stress and strain respectively is shown in Fig. 2.2. While the development of the true stress versus true strain diagram considers instantaneous sample dimensions (the cross-section area of the tensile sample and the gauge length) as a particular stage of the tensile test for calculating the stress and strain. There is a continuous change in dimensions (increase in length and reduction in cross-section) of the sample during the tensile test, therefore engineering stress-strain curve and true stress-strain curve always differ. The true stress is always greater than the engineering stress while the true strain is lower than the engineering strain (Fig. 2.2). The major deviation in a true stress-strain curve from the engineering stress-strain curve is observed beyond the yield point.

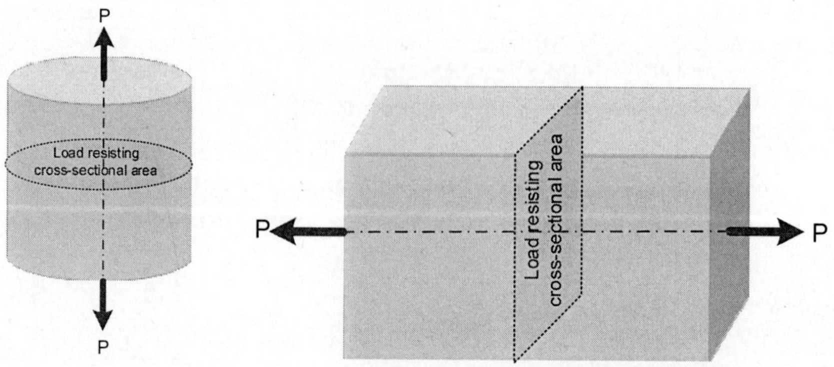

a) Circular and square section

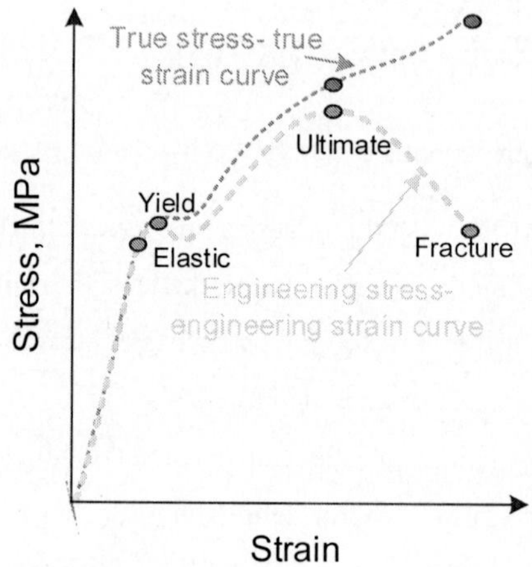

b) Engineering and true stress vs. strain curve

Figure.2.2: Schematic diagram a) tensile loading and relevant (circular and square cross-section) load resisting cross-sectional area and b) engineering and true stress vs. strain relation

2.2.1 Common parameters and calculations

Maximum stress up to which a material shows elastic behavior is termed the elastic limit. The slope (0-1 as marked in Fig 2.1) of the engineering stress-strain curve in the elastic region shows the modulus of elasticity. In this region, the strain is directly proportional to the stress applied. This is referred to as **Hook's law**. The proportionality constant in Hooks law is basically the modulus of elasticity of the material. The application of the higher tensile stress (than the elastic limit) causes plastic deformation. The minimum stress at which a material plastically deforms is known as the yield strength. The plastic deformation of the material causes strain-hardening of materials. Therefore, strain hardening results in increasing stress required to cause further plastic deformation beyond the yield point. The slope of the engineering stress-strain curve in the plastic zone shows how rapidly a metal is being strain hardened due to plastic deformation. Few metals have low stacking fault energy strain hardens faster to a greater extent than other metals of high stacking fault energy like aluminium magnesium. Strain hardening behavior is characterized by a parameter called work hardening exponent (n).

Engineering stress is obtained from the ratio of the tensile load (corresponding to elastic limit, yield point, and ultimate strength point) and the original cross-sectional area of the tensile sample. Engineering strain is calculated using the ratio of change in length to the original length of the tensile sample. Engineering strain is obtained from the ratio of change in page length of the fractured sample (kept in abutting position) to the original gauge length. The area engineering stress strain curve indicates the energy consumed in causing the fracture of the desired sample, which indirectly shows the toughness of the material. The combination of the yield strength and ductility of the material determines the toughness as suggested by the engineering stress-strain curve obtained from the tensile test (Fig. 2.3). High strength and ductility are typical indicators of brittle material while low yield strength and high ductility is shown by soft and ductile materials.

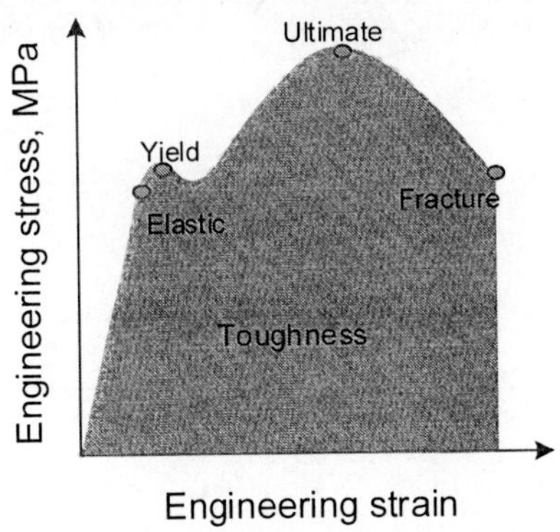

Figure 2.3: Schematic diagram showing the engineering stress vs. strain relation and toughness

Tensile test is conducted using standard sample as per relevant applicable standard. The tensile test sample can be circular or rectangular section usually has three parts namely griping section, graduation cross reduction and gauge length. Gauge length is marked on the sample and gauge size (width, thickness, and diameter) is measured which is used to calculate the stress and strain corresponding to different loading vs. extension data generated from the tensile test. The test is conducted on a tensile or universal testing machine. The sample is firmly secured between the griping jaws of the machine. Then tensile load is gradually increased by moving gripping jaws apart. The speed of jaw determined the rate of increasing tensile load and strain rate. The jaw speed (0.0001 to 1000 mm/min) significantly affects the tensile properties. Increasing strain rate, tends the work harden the metal work piece rapidly which in turn increases the tensile strength and reduces the ductility. Jaw speed and gauge length of the sample is used to calculate the strain rate. The low jaw speed (0.1 to 1 mm/min) is used for hard metals like hardened steel and cast iron); while high speed is preferred for soft metals like Al etc.

Using original cross-sectional area and dimensions of sample the corresponding engineering stresses are obtained based on the elastic point, yield point, ultimate point and fracture point loads (Fig. 2.4). Similarly, using original gauge length of the sample and extension at elastic, yield point, ultimate point and fracture point corresponding engineering strains are obtained. The rest report in addition to modulus of elasticity, engineering stress and strain elastic, yield point, ultimate point and fracture point must include the information related to jaw speed and fracture location. The stages of sample extension is shown in Fig. 2.4 as given below:

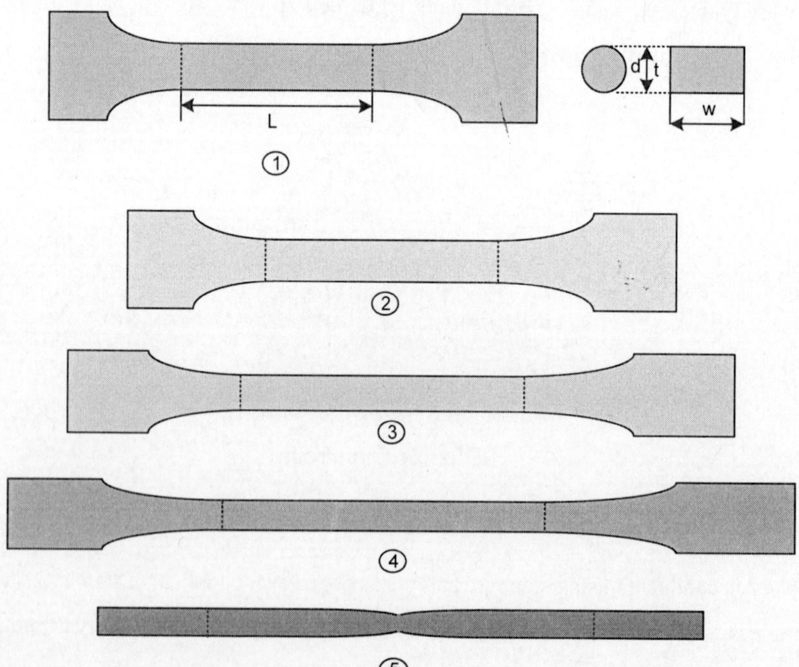

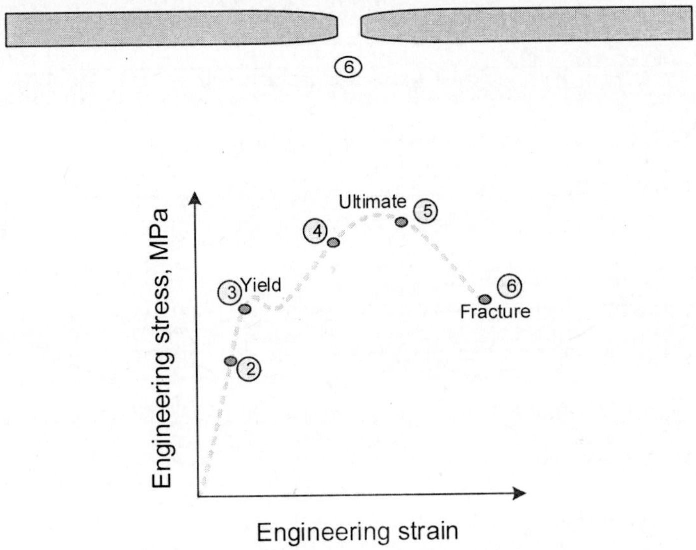

Figure 2.4: Schematic diagram showing a) progressive stages of sample elongation during the tensile test sample until fracture and corresponding points shown in a) engineering stress vs strain diagram.

2.2.2 True stress and true strain

The reduction in cross-section and increase in length of the tensile sample occur gradually and uniformly in the entire gauge length up to ultimate point load with increase of tensile load during the tensile test. The work hardening of the tensile sample depending up on the metal system and jaw speed / rate of loading / strain rate is observed beyond the yield point and continues till fracture due to increasing strain/plastic deformation as shown in Fig. 2.5. The localized reduction in cross section the tensile sample begins at the ultimate point, which is called necking. The necking increases the true stress rapidly after ultimate point due to reduction in actual load resisting cross-sectional area. On the other, necking decreases the load requirement to cause further plastic strain beyond the ultimate point. Use of original load resisting cross-sectional area and reduction in tensile load (requirement for further plastic deformation during test) after ultimate point therefore shows decreasing / negative trend engineering stress with increasing strain.

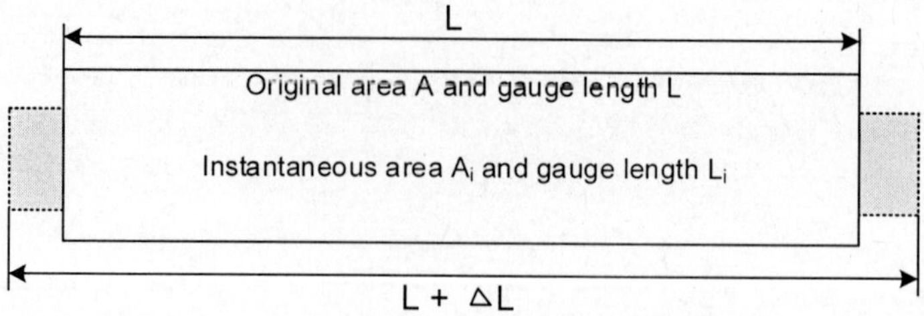

Figure 2.5: Schematic diagram showing increase in length (elongation in tensile loading direction) and decrease in cross-section (lateral contraction) during tensile test.

There is no change in volume of metal in gauge length during the tensile test. Considering volume constancy situation:

Original Cross-sectional Area, A X Original Gauge Length, L = Instantaneous Cross-sectional Area, A_i X Instantaneous Gauge Length, L_i

$Ai = A X L / Li$

At any stage of deformation during the test Li: $L + \Delta L$ (change in length)

Engineering stress σ : Load, P / Original cross-sectional, $A = P/A$

True stress: Load, P / Instantaneous Cross-sectional Area, $A_i = P / A_i$

True stress : $P/ A_i = P/ (A.L / Li) = (P/A).(Li/L) = \sigma . [(L + \Delta L)/L] = \sigma . [(L/L) + (\Delta L/L)]$

True stress : $\sigma . [1 + \varepsilon]$ = **Engineering stress (1 + Engineering strain)**

True strain: ln.$(Li/L) = ln.[(L + \Delta L)/L] = ln.[(L/L) + (\Delta L/L)]$

True strain: ln. $(1+ \varepsilon)$ = **(1 + Engineering strain)**

2.2.3 Yielding and yield strength

The materials under the influence of external load generally first show the (reversible) elastic deformation within the elastic limit. Further, increase of external load/stress beyond the elastic limit results in (permanent) plastic deformation. Within the elastic limit strain is found to be directly proportional to stress applied which in commonly known as Hooks law. Stress and stress (within the elastic limit) relationship is perfectly linear. The proportionality constant relating to the stress and strain is a materials properties known as modulus of elasticity and modulus of rigidity in case of tensile / compressive and shear stress respectively. The ratio of tensile/compressive stress and the corresponding strain gives us modulus of elasticity €, while ratio of shear stress and corresponding strain gives us modulus of rigidity (G). The yielding refers to (commencement of) plastic deformation. The stress level at which commencement of yielding or permanent plastic deformation of a metal occurs is called yield strength. Few metals like carbon steel show very clear deviation from linear stress vs strain (load vs. Extension curve) plot suggesting the specific yield strength and commencement of plastic deformation (Fig. 2.6). Many metals like Al, austenitic stainless steel etc. do not exhibit clear deviation from linear stress vs. strain (load vs. extension) relation in such cases determination of commencement of yields and yield strength becomes little difficult and different approach is used to establish the same.

As practice, first a stress vs strain plot of such metals is developed by conducting a tensile test. A line beginning either from 0.002 or 0.005 strain is drawn parallel to straight line showing linear relation between stress and strain (within elastic limit). 0.2% and 0.5% strain is considered for most of metals to determine the 0.2% and 0.5% off-set yield strength. The point of intersection of parallel line (drawn from 0.002 or 005 strain as per case) with stress vs. strain curve shows the 0.2% yield point or 0.5% yield point and corresponding stress value shows yield strength of metal accordingly (Fig.2.6).

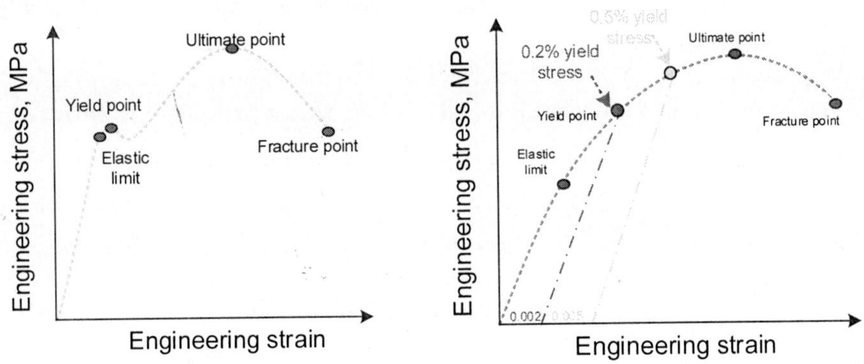

a) b)

Figure 2.6: Schematic diagram showing how to establish yield point / strength of a) metals like steel and b) offset yield strength of metals like Al, austenitic stainless steel

2.2.4 Ductility

The ductility shows how far a material can be elongated / deformed plastically prior to fracture. The ductility of a material is a mechanical property of great importance from the manufacturing and defect tolerance points of view. A ductile material resists the nucleation and growth of a crack, thereby making it more fracture resistant especially in presence of a stress raiser like defect, geometrical notches etc. The metal must be ductile enough for having reasonable ease of manufacturing using bulk deformation based manufacturing processes like forging and rolling etc. The ductility of a metal can be evaluated by conducting the tensile test and measuring either % elongation or % reduction in area at the fracture. High % elongation (strain) and % reduction in cross-sectional area shows high ductility of the metal. The % elongation is obtained from the % age change (increase) in gauge length (till fracture) with respect to original gauge length. Change in gauge length of the sample after fracture is obtained by putting the sample in abutting position and measuring increase of gauge length (Fig. 2.7).

$$\%\ Elongation\text{: } (\Delta L \div L)100$$

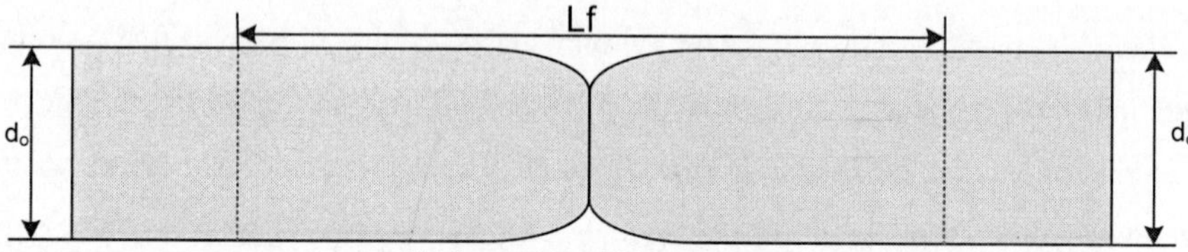

Figure 2.7: Schematic diagram showing the way to determine the final length of sample after fracture to obtain the elongation

The % reduction in area is obtained from the % age change (reduction) in cross-section of fractured sample surface with respect to original cross-sectional area. The reduction in cross-sectional of the tensile sample after fracture is obtained by measuring area of fracture sample surface (average of two sides) with respect to original cross-sectional area (Fig. 2.8).

$$\%\ \boldsymbol{Reduction\ in\ area}\text{: } \boldsymbol{(\Delta A/A)100}$$

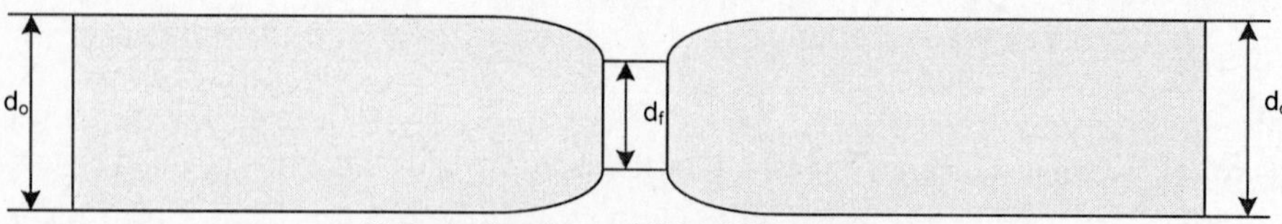

Figure2.8: Schematic diagram showing the way to determine the final cross-sectional area after fracture of sample to obtain % elongation

2.2.5 Resilience

The resilience of material is the elastic strain energy stored (U). It depends on the yield stress and modulus of elasticity of the materials. The area under the engineering stress-strain curve up to the elastic limit shows the resilience of the material (Fig. 2.9). The resilience is obtained using the relationship 1/2 [(elastic stress).elastic strain]. Resilience indicates the amount of energy a material can absorb within the elastic limit (prior to the commencement of the yielding) and release the same on the removal of load and regain the shape and size. It is assumed that some amount of the strain (0.002) is left in the material even after unloading the sample. This is taken into account while considering the elastic strain.

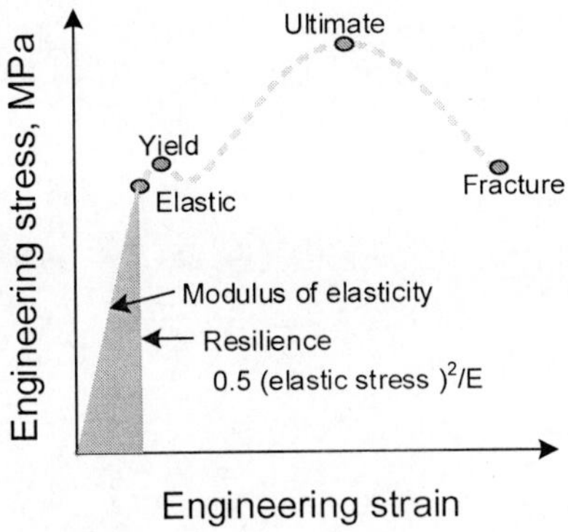

Figure 2.9: Schematic diagram showing resilience and modulus elasticity obtained with help of engineering stress vs strain diagram.

2.2.6 Elastic modulus

The elastic modulus also called modulus of elasticity (E) of a material indicates rigidity and is obtained from the ratio of stress to strain. For a given applied load / stress, the modulus of elasticity (E) can be used to calculate strain (σ / E) within the elastic limit. Similarly, the modulus of elasticity (E) can also be used to calculate the residual stress (ε / E) using locked strain obtained experimentally. High modulus of elasticity (E) of a material means low elastic strain for a given applied stress. The relationship between modulus of elasticity (E) and modulus of rigidity (G) is expressed using following: E =2G (1+ μ). Where μ is the poison ratio which a ratio of transverse (lateral), ε_t to longitudinal, ε_l (axial) strain. It is negative ratio and value mostly ranges from 0.3 to 0.6.

2.3 Compression test

The compression test is primarily used to study behaviour of material under crushing load conditions. The test is conducted in a way similar to that of the tensile test by applying compressive pressure on the sample to establish the elastic limit, modulus of elasticity, proportionality limit, yield strength and ultimate compressive strength (Fig. 2.10). The compression test helps to evaluate capability, integrity and suitability of a material under compression during deformation based manufacturing, and service like in construction sector.

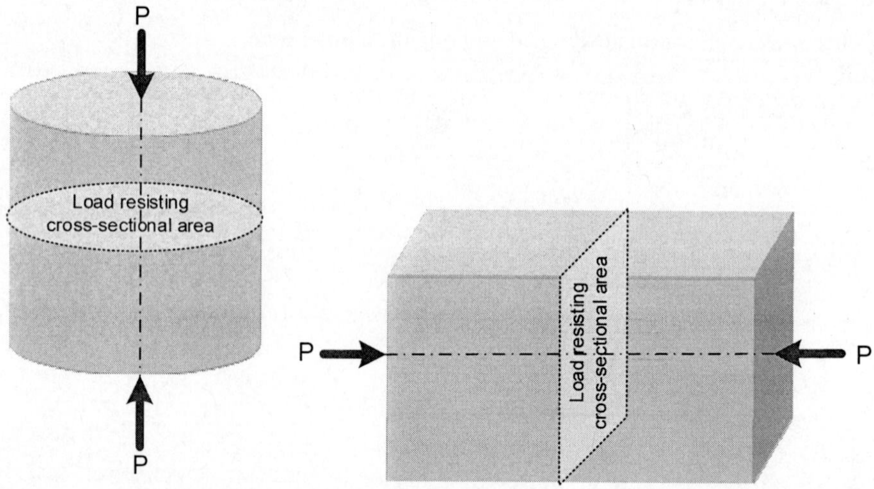

Figure 2.10: Schematic diagram showing compressive loading and relevant (circular and square cross-section) load resisting cross-sectional area

Generally, materials show inverse relationship between tensile strength and compressive strength. Conversely, high compressive strength materials (like concrete, ceramics, composites, cardboard etc.) show low tensile strength materials. Metal under compression behaves almost similar to that of tension (Fig. 2.11). Corrugated materials like porous metals and cardboard etc. show completely different behaviour as shown in Fig. 2.12. The compression test usually performed on the finished products such as windshield, plastic furniture and pipe, balls used in games like tennis, golf. Idea behind consideration of compressive strength properties is in design of such products that strike a good balance with mechanical capability of product and the material use.

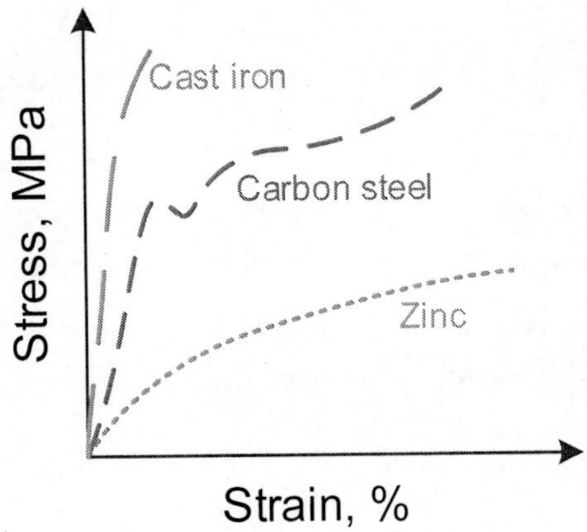

Figure 2.11: Schematic diagram showing engineering stress vs strain diagram under compressive loading for different metals

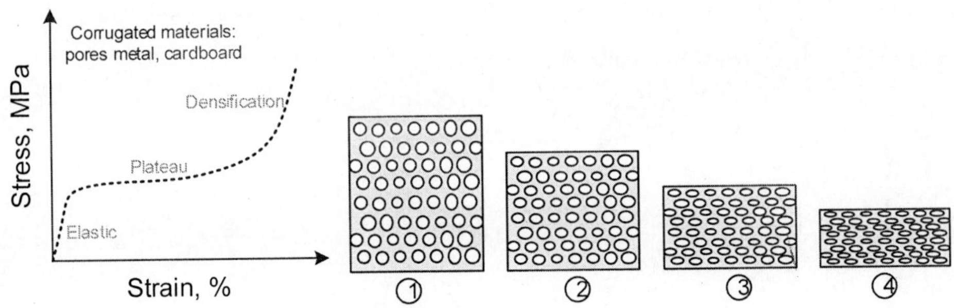

Figure 2.12: Schematic diagram showing engineering stress vs strain diagram under compressive loading for porous metals

During the compression test, the sample material to be tested is kept between the gripping jaws of universal testing machine and compression load is increased gradually (by moving jaws close to each other at different speed e.g. 0.001 to 100 mm/min) and the load-displacement curve is obtained. Based on the load-displacement curve data corresponding compressive stress-strain curve is plotted which gives us information (like elastic limit, modulus of elasticity, proportionality limit, yield strength and ultimate compressive strength etc.) used for design purpose. The materials under compression can fail in different ways namely brittle fracture, barrelling and buckling as shown in Fig. 2.13.

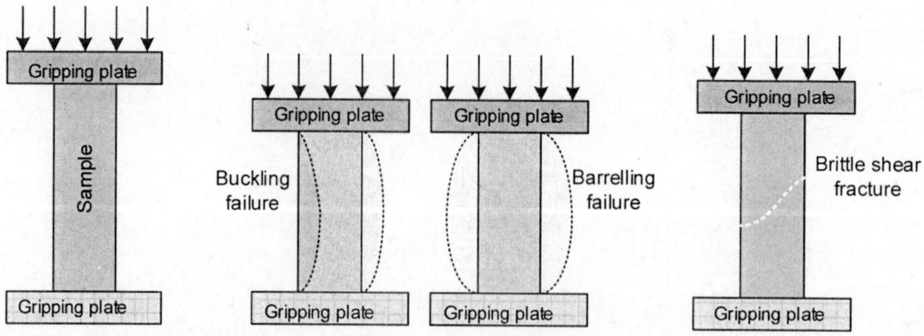

Fig. 2.13 Schematic diagram showing different modes of failure under compressive loading

The trend of stress-strain curve definitely depends on the material characteristics, as it is evident from schematic diagram showing stress-strain curve relationship for ductile and brittle materials. The ductile material (like thermo-plastics) do not fracture but continue to yield until the compressive load in increased while in case of brittle materials (like ceramics, concrete) after attaining the maxima (ultimate compressive stress) load decreases sharply as materials is crushed (Fig. 2.14). In former case, the compressive strength is obtained based on 2, 5, 10% reduction in gauge length of compression sample.

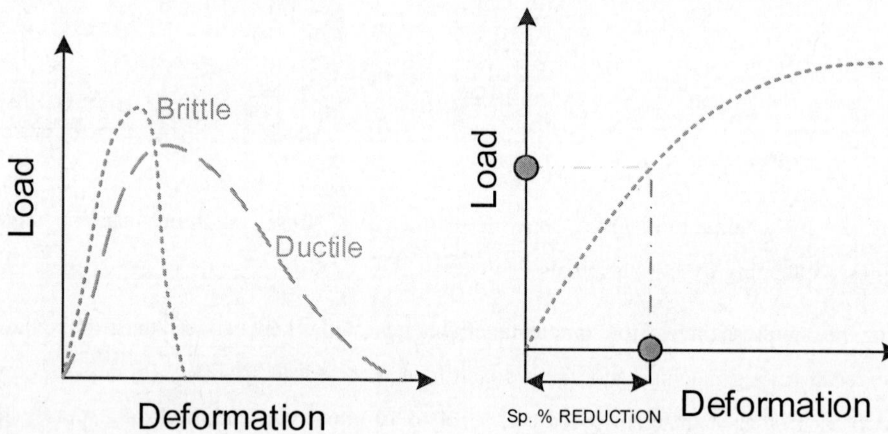

Figure 2.14: Schematic diagram showing a) load-deformation relation for ductile and brittle materials and b) methodology to obtain yield point under compressive loading

2.4 Torsion test

Torsion is the twisting of components by the moment acting along its axis. Many engineering components like shafts, wrenches are subjected to such twisting moment to cause rotation. The twisting moment obtained from the product of tangential force (F) acting and its distance (r) from axis of the component called torque (F.r in Nm). The torque tends to cause rotational deformation in the component.

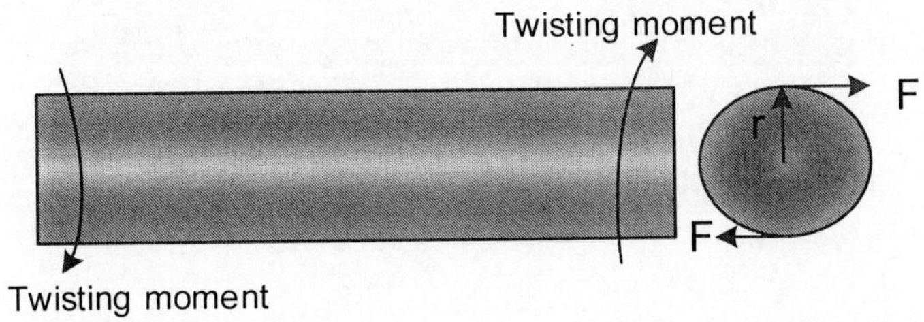

Figure 2.15: Schematic diagram showing torsional loading of the circular cross-section sample

The rotational deformation under influence of torque is measured in terms of angle of twist (Θ). For successful performance many component the angle of twist per unit length must minimum. Consider the following assumptions for torsional analysis. Consider circular section sample subjected to the torque T at its ends. A rotation at one end of the bar relative to the other end will occur (Fig. 2.15). The rotation angle of the circular section (angle of twist) is Φ. The longitudinal distortion along the shaft at angle is γ.

- A plane section of circular section sample perpendicular to the axis remains plane even after the application of the torque
- Under the influence of torque, shearing strain in circular section varies linearly from the axis.
- Shear stress is proportional to shear strain.

The angle of twist can be calculated using following equation:

Angle of twist, Θ: $T.L/G.J$

Where T is the torque applied (N. m), length of the longitudinal axis of component (m), G is modulus of rigidity (N/m^2), J is the polar moment of inertia (m^2). Thus, the above equation suggests that the angular deformation (angle of twist) under twisting moment depends on torque applied, length of component under consideration, modulus of rigidity (material property) and polar moment of inertia as per cross-section of the component (Fig. 2.16). Increasing torque and length of component (distance from one (fixed) end where twisting moment applied and decreasing modulus of rigidity and cross-section area in general increases the angle of twisting.

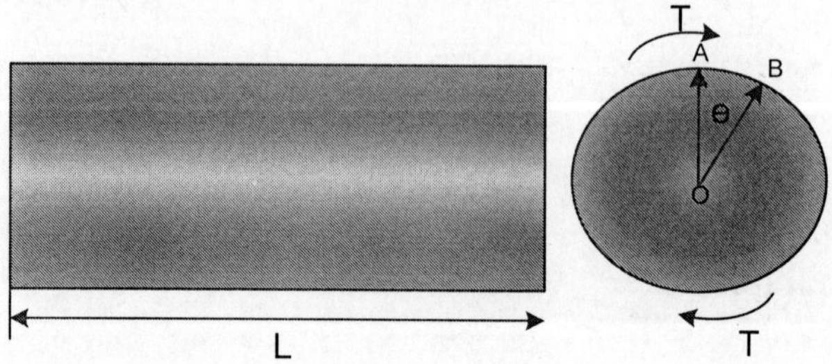

Figure 2.16: Schematic diagram showing twisting due to torsional loading of the circular cross-section sample

The polar moment of inertia indicates the resistance of component to rotation deformation under twisting moment due to its geometry and cross-sectional area (Fig. 2.17). For circular hollow and solid section, the polar moment of inertia obtained using the following relation.

The polar moment of inertia of hollow circular section, J: $[\pi (r_o^4 - r_i^4)]/2$

The polar moment of inertia of solid circular section, J: $[\pi (r^4)]/2$

(as r_i becomes zero, r_o becomes r as radius of the solid cross-section)

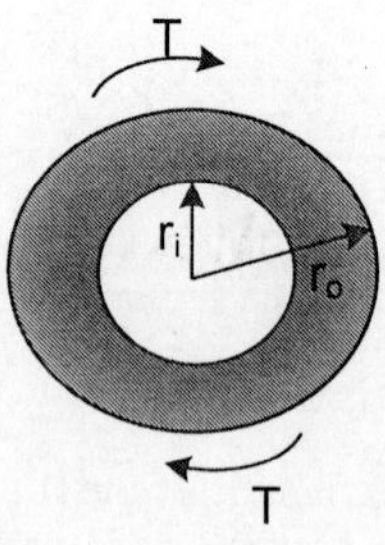

Figure 2.17: Schematic diagram showing torsional loading (torque) on the hollow circular cross-section sample

Further the angle of twist for given length of sample and torque applied can be used to calculate the modulus of rigidity G: $T.L/ \Theta.J$

The torsion induces shear stress and shear strain in the component. The shear strain (λ) at the surface of the component due to the torque can be obtained from (*RQ/PQ: rΘ/L*). The shear strain equation suggests that distance of the layer / materials away from centre (r), length of the component for a given angular twist directly affects the shear strain induced. The sear strain at the centre of section (where r is 0) will be absent. Conversely, shear strain increases linearly with distance from the centre/axis of the object.

The shear stress induced in a component due to torque increases linearly from centre to the surface causing maximum stress at the surface (Fig. 2.18). The hollow cylindrical section are considered to be more effective for torsional loading as very low stresses are induced at the internal section near the centre. The shear stress (τ) at a location (distance r from the centre) can be obtained from the following relation: τ*: (T.r)/J*. This equation suggests that increasing torque and distance from centre and reducing polar moment of inertia directly increase the shear stress.

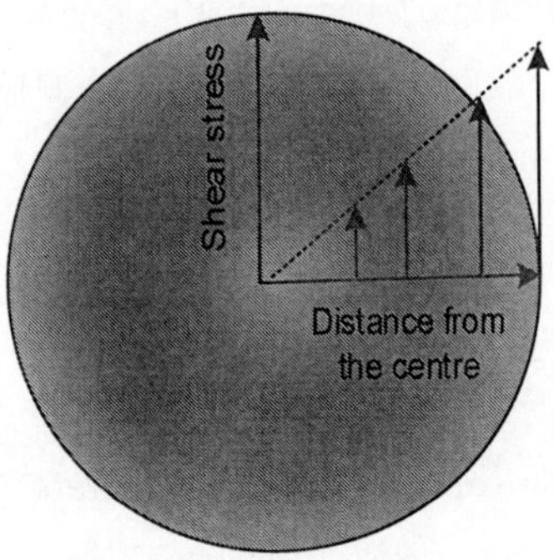

Figure 2.18: Schematic diagram showing shear stress distribution (from the centre to outer periphery) on circular section under torsional loading (torque)

The suitability of material under the twisting load conditions for the applications like power transmission using a shaft is examined by torsion test, which shows the resistance of the material under the shear stress induced by the twisting moment. The torsion test induces the shear stress in the material. The torque vs. angular twist relationship is obtained to establish the elastic limit, yield stress, ultimate stress, and fracture point like a tensile test (Fig. 2.19). Moreover, it is required to calculate shear stress induced by a torque/ twisting moment and shear strain induced in terms of angular twist achieved. Shear stress calculation considers the shear area of a component and shear stress is angular twist per limit length.

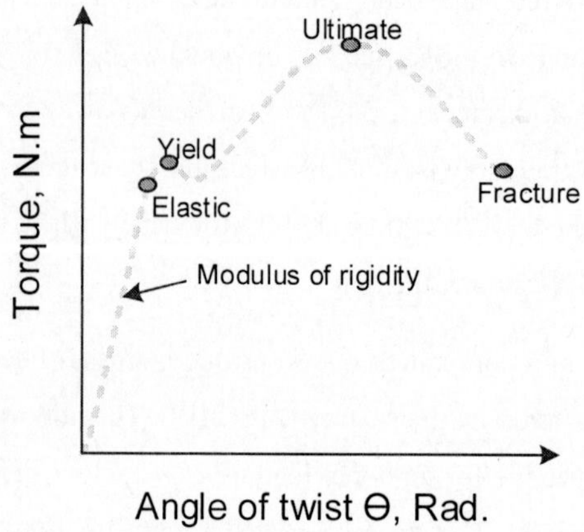

Figure 2.19: Schematic diagram showing torque (torsional loading) vs. angle of twist relation

The cross-section component can be circular, flat, or any other shape as per the application. In light of the requirement of the application, the material can be subject to static or fatigue conditions, and accordingly, a suitable torsional test is designed. It can be done to establish the shear stress-shear strain behaviour, proof stress, ability to withstand the fatigue load conditions, or functional suitability of the material as per need.

The behavior of materials under static, torsional loading conditions is examined by continuously increasing the torque in one direction and measuring the corresponding angle of the twist of the sample. Using the torque versus twisting angle relationship, shear stress vs shear strain curve is developed to obtain the various parameters like elastic limit, yield point, ultimate point, and rupture point (Fig. 2.20). Resilience under the torsional load conditions of material can be identified from the area and under shear stress vs strain curve within the elastic limit. Similarly, the energy consumed in causing the fracture under torsional conditions is obtained from the area under the shear stress vs. strain curve till the fracture point.

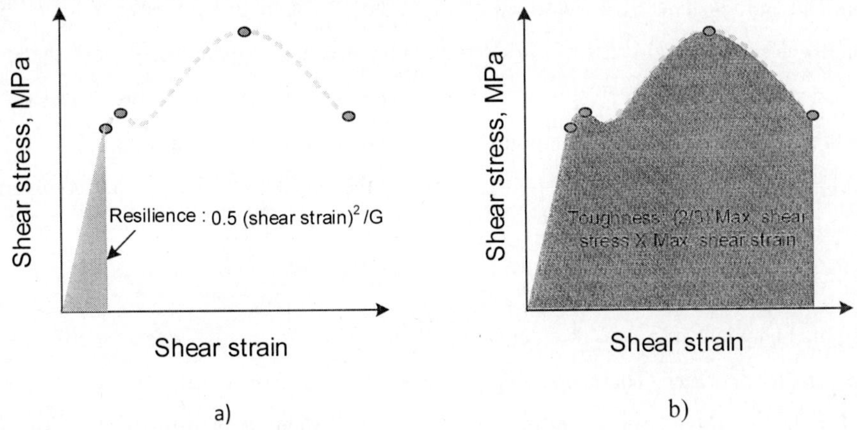

Figure 2.20: Schematic diagram showing a) resilience and b) toughness of material under torsional loading with help of shear stress vs strain diagram.

The fatigue behaviour of the material under torsional load conditions is evaluated by applying the twisting moment in clock and anticlockwise directions to induce the reversible shear stresses and the number of cycles to cause the fracture is identified. A number of cycles of torsional fatigue loading for which a specimen withstands indicate the fatigue resistance. The torsional behaviour of the material can also be studied by applying unique torsional loading or twisting movement as per the need in light of the surface conditions for successful application.

2.5 Toughness

The toughness of a material shows the impact resistance. It is determined by impact toughness tests namely Izod and Charpy test. The energy consumed by a standard sample during toughness test is used as a measure of toughness of the material. Higher the impact energy absorbed by the materials (to cause fracture/ deformation-fracture) during a toughness test, greater is the toughness of material. As described earlier, tensile test also shows toughness as a combination strength and ductility, which is the energy consumed to cause the fracture of tensile sample under (almost static loading) as load is increased very gradually. While in case of the impact toughness test, (impact) load is applied at very high rate. The impact toughness value (energy absorbed in J or Nm) is not directly used in mechanical design calculations. Rather it indicates the behaviour of materials under real life-impact load conditions.

In actual practice, engineering components are subjected to various kinds of loads namely static and dynamic. This classification is based on the rate of change in magnitude of load and direction. Dynamic loads are characterized by high rate of change in load magnitude and the direction. Reverse happens in case of static loads. In the hardness test and tensile test, the load is increased very slowly that corresponds to the behaviour of material under more or less static loads. Moreover, very wide range of loading rate applied by moving sample holding jaws apart at different speeds (i.e. 0.0001 to 1000 mm/min) during the tensile test. Rate of loading governs the strain rate and therefore behaviour of materials. At low rate of loading a material showing the ductile behaviour may respond like brittle materials under high rate of loading conditions.

To study the behaviour of material under impact load conditions at high rate of loading, and toughness test is conducted. There are two methods used for toughness testing namely Izod and Charpy test, based on the common principle of applying the load at high rate and measuring the amount of energy absorbed (N. m or Joule) in breaking the sample due to impact. However, these two methods differ in terms of standard sample (i.e. geometry), sample holding method and energy of the heavy weight pendulum that hits the sample (Fig. 2.21).

Sr. No.	Toughness test	SampleHolding
1	Izod	Cantilever and notch faces the pendulum
2	Charpy	Simply supported and notch is opposite side of the pendulum

Standard sample for both testing methods has a notch due to reason explained below. The most of the engineering components are invariably designed with some kind of notch and stress raisers. It becomes important to see the behaviour of material under impact loading in notched condition. Therefore, toughness test is conducted using a sample having a notch. Moreover, un-notched samples can also be used for the test using non-standard sample and then results are expressed accordingly. The location and direction (i.e. with respect rolling) wherefrom the sample has been taken affecting the toughness significantly. Therefore, both location and direction of sample where notch is made must be reported with toughness test results.

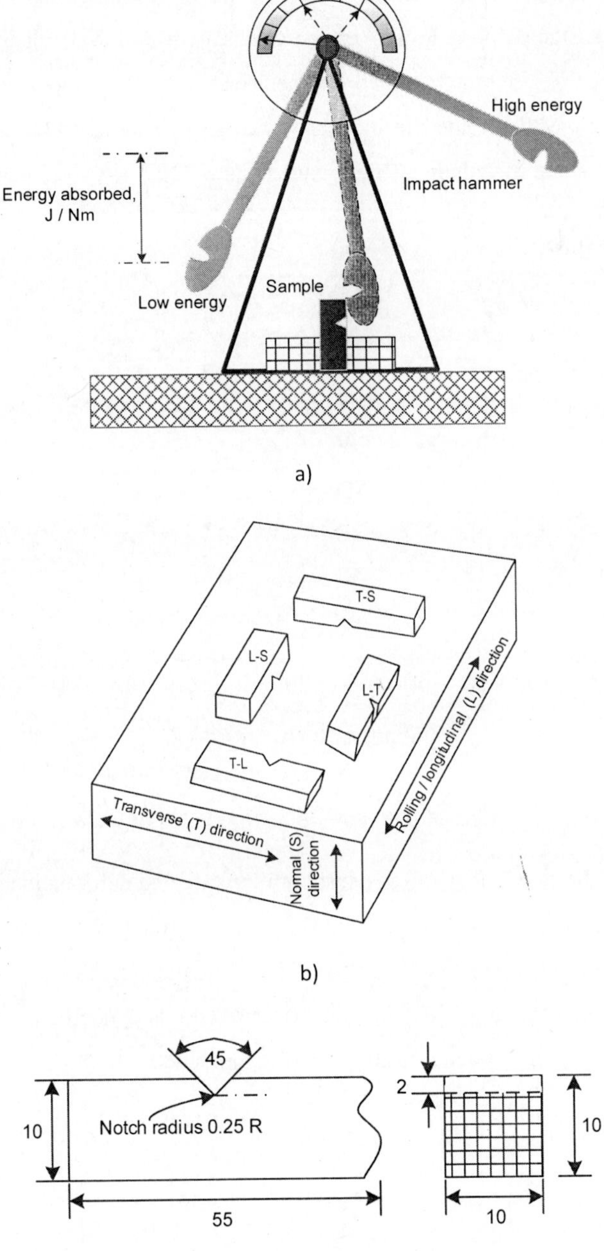

Figure 2.21: Schematic diagram showing a) principle of impact toughness test and b) identification of location and direction for toughness sample collection and c) standard sample with details of notch.

Result of toughness test is expressed in terms of amount of energy absorbed, which is not directly used for design purpose. Moreover, these tests are useful for comparing the resistance to impact loading of different materials or materials of same composition with different conditions of heat treatment and mechanical working. Resistance to the impact loading of material depends on the surrounding temperature (Fig. 2.22). Therefore, temperature at which toughness test is conducted must be recorded with results.

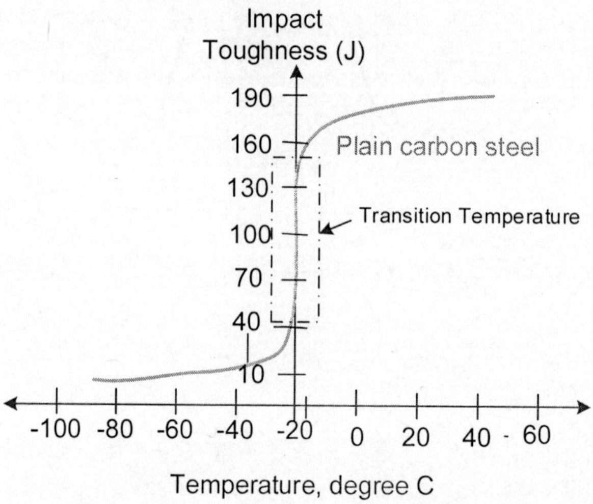

Figure 2.22: Plot showing effect of temperature on toughness of plain carbon steel

Toughness is the ability of a material to resist both fracture and deformation. The fracture of a brittle material after impact toughness testing shows flat & smooth surface while that of ductile materials show deformation coupled with significant lateral contraction as shown in Fig. 2.23. Tough materials show a good combination of strength and ductile. Notches in impact specimens increase the stress concentration and fracture tendency.

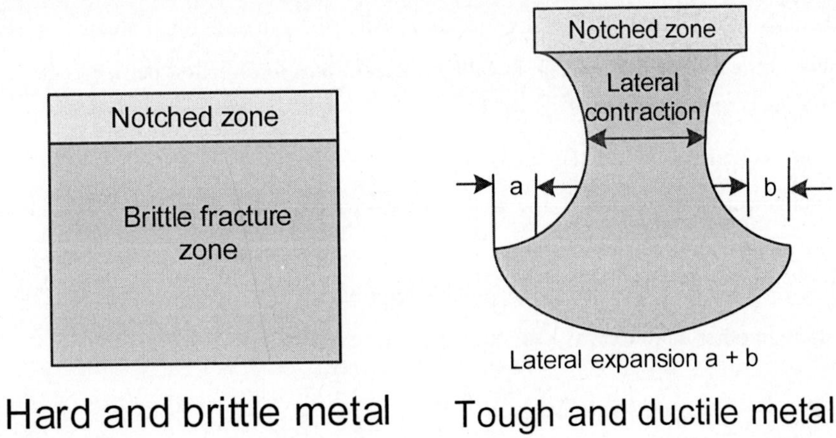

Figure 2.23: Schematic diagram showing fracture-surface after toughness test of sample made of a) brittle material and b) ductile material

2.6 Elastic recovery

A material when subjected to loading / stress beyond the yield point experiences first the elastic strain then plastic strain. Removal of the load from any stage after inducing plastic strain results in minor reduction in the strain observed at load / stress removal point (PS+ ES) as shown in Figure. This primarily occurs due to disappearance of the elastic strain (ES). The phenomenon is termed as elastic recovery. On reloading the same sample further plastic strain is caused from the same point (of strain) at which unloading was done. The elastic recovery is considered to be around 0.002 % strain (i.e. 0.2 strain).

The elastic recovery is of greater practical importance from the manufacturing point of view especially in case of deformation based manufacturing processes. The elastic recovery makes it difficult to realize the desired size and shape. It must be considered while design a deformation based manufacturing process so that metal is plastically deformed little extra (equal to elastic strain) so that after removal of a load / stress applied during manufacturing the desired size and shape achieved in the manufactured product.

2.7 Generalized Hook's law

Stress is directly proportional to strain within the elastic limit and is expressed by following relation: Stress: strain and elastic modulus [$\sigma = E.\varepsilon$]. This law hold good for uni-axial stress in elastic zone only and can be used to calculate the elastic strain induced by stress [$\varepsilon = \sigma/E$]. Application of tensile stress causes tensile strain (longitudinal strain) my elongated the metal along the axis of applied load. Metal under the tensile stress in additional elongation always experiences little contraction (which is called lateral strain) in direction perpendicular to the axis of tensile load (Fig. 2.24). The ratio of lateral (ε_{lat}) to longitudinal strain (ε_{long}) within the elastic limits called poison ratio ($v = \varepsilon_{long}/\varepsilon_{long}$).

In many engineering applications, the metals are exposed to bi-axial and tri-axial loading as well. The deformation behaviour of a metal under such conditions can obtained using generalized Hook's law. This law considers the deformation in a particular direction due to all stress acting in other directions as well but within the elastic limit.

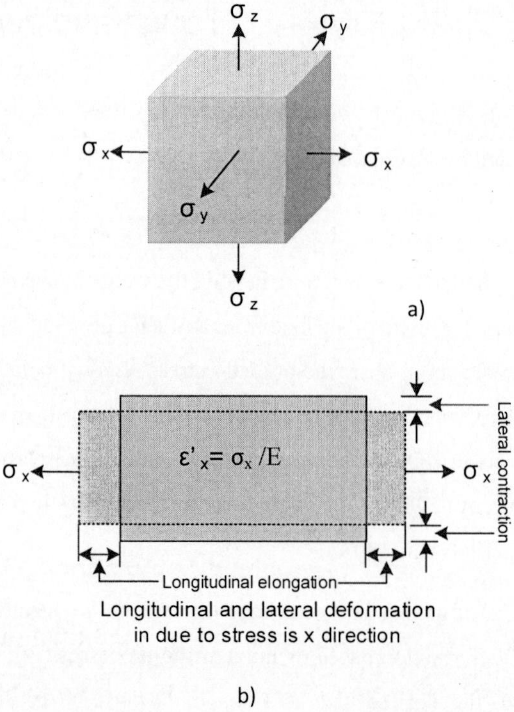

Figure 2.24: Schematic diagram showing a) cuboid subjected to tri-axial stresses (x, y, z directions) and b) longitudinal elongation and lateral contraction due to stress acting in direction x

Let's consider case where in a cuboid is subjected to tri-axial stresses ($\sigma_x, \sigma_y, \sigma_z$) due to tensile load in three directions *x, y, z* causing corresponding three strains (ε_x , ε_y , ε_z). The longitudinal strain due to loading in x direction can be obtained from ($\varepsilon'_x = \sigma_x / E$). However, metal (in direction x) will also be experiencing lateral strains (ε''_x and ε'''_x) also due to stresses (σ_y, σ_z)

other two directions as well (Fig. 2.25). The lateral strain is considered negative as contraction takes place while longitudinal strain is positive where in elongation takes place.

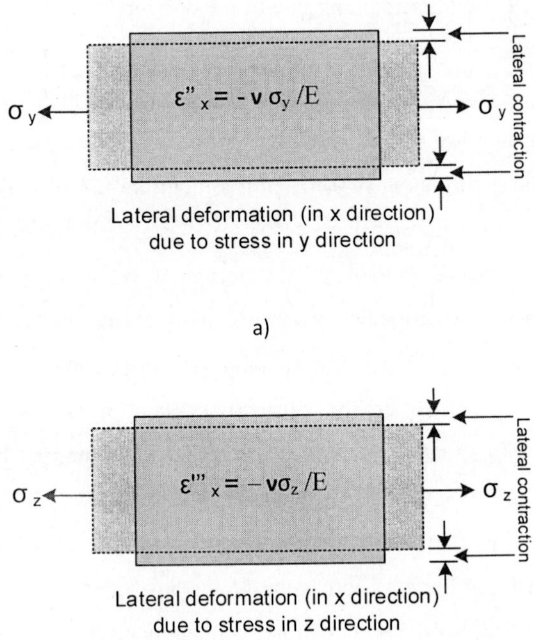

b)

Figure 2.25: Schematic diagram showing lateral contraction in direction x due to stress acting in direction a) y and z

The lateral strain in direction x due to stress (σ_y) will be $\varepsilon''_x = -\nu\,\sigma_y/E$. Similarly, the lateral strain in direction x due to stress (σ_z) will be $\varepsilon'''_x = -\nu\,\sigma_z/E$. Thus, total strain (longitudinal and laterals) in direction x due to all stresses (σ_x, σ_y, σ_z) will be:

$$\varepsilon_x = (\varepsilon'_x + \varepsilon'_x + \varepsilon''_x) = (\sigma_x/E - \nu\,\sigma_y/E - \nu\sigma_z/E) = (\sigma_x/E - \nu\,\sigma_y/E - \nu\sigma_z/E)$$

Total strain in direction x: $\varepsilon_x = \frac{1}{E}[\sigma x - \nu\,(\sigma y + \sigma z)$

Similarly, total strain in direction y: $\varepsilon_y = \frac{1}{E}[\sigma y - \nu\,(\sigma x + \sigma z)$

Total strain in direction z: $\varepsilon_z = \frac{1}{E}[\sigma z - \nu\,(\sigma x + \sigma y)$

2.8 Hardness Test

The application of load on the surface of a material through sharp and pointed objects frequently causes penetration, scratching due to relative movement. Surface of the materials must resist such kind penetration / indentation, which under relative motions conditions causes wear and roughens the surfaces. The property of a material, which shows the resistance to indentation, and scratching called hardness. Since an indentation occurs through the displacement / plastic deformation of a material around the indenter, therefore resistance to indentation (hardness) also indicates the resistance of plastic deformation or yielding. Hardness is very extensively used material characteristics in engineering application. Hardness of metals directly related with tensile strength. Higher the hardness greater is tensile strength. Soft material shows low hardness. However, like toughness, the value of hardness is also directly not used in mechanical design of components and systems. The hardness measurement based on load and indenter is broadly grouped as macro-hardness (normal hardness) and micro-hardness. Very fine size of inventor (micrometres) and very low load (in grams) is used for micro-hardness test. The fine indenter allows the measurement of hardness of individual grains/constituents and hardness of across the grains and zones (like across the weld joint, across the coating interfaces etc.). On contrary, conventional/ macro/ normal hardness uses a large size indenter (i.e. from few mm to tens of mm) and high load (from 100 to 3000 kg) covering a large surface area including hundreds to thousands of grains/crystals. The normal hardness give a more representative / average value of surface hardness. Therefore, the micro-hardness suits better for research and development while normal hardness is more useful for industrial application / characterization.

Moreover, hardness measurement gives many useful information and inferences as listed below:

- The hardness shows extent to which service conditions (e.g. due to surface layer hardening, chemical compound formation, carburizing/decarburizing, oxidation etc.) affected the component
- The hardness allows estimating the tensile strength of material indirectly besides indirectly giving idea about tough and ductility
- Hardness profile of a component, weld joint , coating etc. shows the weak location (minima hardness zone) which needs attention to avoid service failures

- Hardness data can be used for methods of manufacturing of component of a particular metal
- Hardness shows how far bulk or surface property modification processes (like heat treatment, casehardening coatings etc.) are effective.
- Hardness measured is very crucial for failure analysis and prevention of metallic components

There are many methods of hardness measurement namely Brinnel, Rockwell, Vickers, Knoop etc. All these methods are based on the same principle of indenting the surface of sample (i.e. for hardness test) using appropriate indenter, and load followed by measurement of indentation accordingly. Thereafter, evaluating/ measuring the features indentations like depth of penetration/ indentation, diameter of indentation at the surface, average length of diagonal of indentation at the surface of the sample. Greater the effect (i.e. in terms of diameter, depth, diagonal of indentation etc.) of given indenter (for a given load) lower is hardness.

Table 2.1 Comparative technical aspects of different hardness test methods

Parameters	Brinell	Rockwell	Knoop	Vickers
Load	500-3000 kg	Minor load: 10 kg Major load: **Cone (120°):** A 60 kg, C: 150 kg D: 100 (including minor load) as per scale **Ball (1/16 inch)**: B 100 kg	25 g- 5 kg	1-120 kg
Indenter	Ball	Ball or cone	Pyramid	Pyramid
Measurement	Diameter	Depth	Diagonal	Diagonal
Indentation	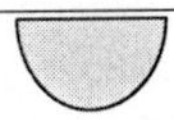	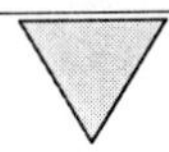	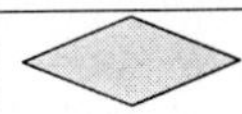	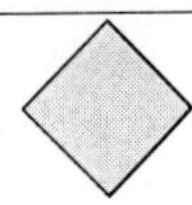
Formulation	$2P/\pi D[D-(D^2-d^2)^{1/2}]$		14.2 F/D^2	1.854 F/d^2
Terms	P load in kg, D indenter diameter in mm, d diameter of indentation in mm		F load in kg, D long diagonal	F load in kg, d average diagonal length in mm
Hardness range	Soft to hard	*B* scale: Soft to medium	Very soft to very hard	Very soft to very hard

		C scale: Medium to very hard		
Surface preparation	Coarse sand	Fine sand	Fine polishing	Polishing

2.8.1 Brinnel Hardness Test

This test is used to measure the hardness of very wide range of materials from soft to hard. Load (500 to 3000 kg) is applied on surface of materials through a steel ball indenter of 10 mm diameter, which produces a hemispherical shape indentation. The diameter of indentation on the sample surface is measured (Fig. 2.26). In general, a large diameter of indentation shows low hardness and vice versa. Using load applied, diameter of indentation and diameter of indentation Brinnel hardness number (BHN) is calculated using equation as given in Table 2.1. Empirically the tensile strength (MPa) of the carbon and alloy steel is found to be about 3.4 times of BHN.

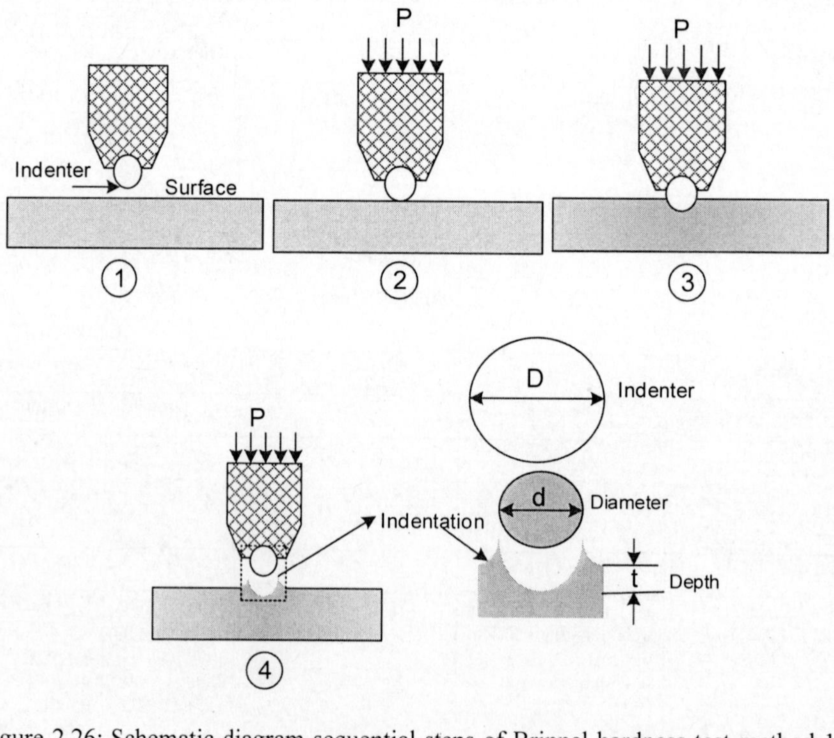

Figure 2.26: Schematic diagram sequential steps of Brinnel hardness test method 1) surface ready for testing, 2) indenter in contact of the surface without load and 3) normal load applied

resulting in desired indentation and 4) indenter retracted showing closure view of indentation and different dimensions

2.8.2 Rockwell Test

This is one of the most commonly used hardness test for soft to hard materials. This method uses different scales as per indenter ball / cone and load applied through indenter as mentioned in Table 2.1. The fine sand polished finished surface of the sample is initially subjected to a minor load to ensure the firm metal surface to indenter contact, thereafter, as per chosen scale (in light of expected hardness of material) major load is applied. The indenter penetrates the surface (Fig. 2.27). The depth of indentation for a given major load (indirectly) indicates the hardness. In general, a deeper penetration (depth of indentation) shows low hardness and vice versa. The following section shows the relationship hardness to tensile Strength (with error +/- 15%):

- For 82 < Rockwell B < 100: Tensile strength (psi): (4750000 - 12000 Rockwell B) / (130 - Rockwell -B)
- For 10 < Rockwell C < 40: Tensile strength (psi): 10^5 (7000 - 10 Rockwell C) / (100 - Rockwell -C)^2

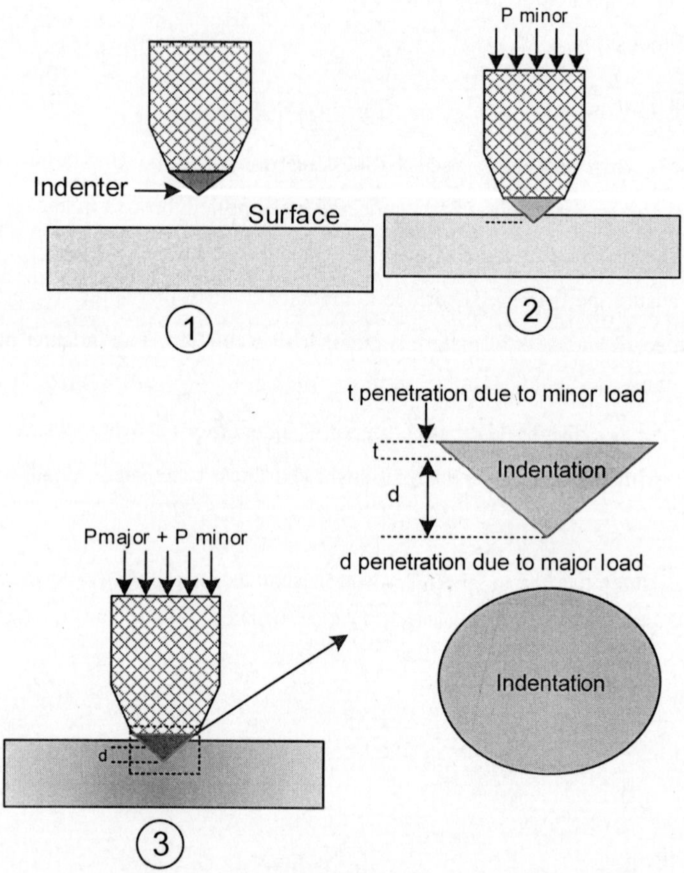

Figure 2.27: Schematic diagram sequential steps of Rockwell hardness test method 1) prepare surface ready for testing, 2) minor load applied through indenter on the surface 3) minor and major load applied through indenter resulting in desired indentation showing closure view of indentation (in top and front view) and different dimensions

2.8.3 Vickers hardness test

This test is used to examine the of hardness fine polished surfaces of small sample of very soft to very hard material using diamond square shaper pyramid indenter with varying load as per sample material. The Vickers hardness test is very accurate and requires less load. The test can be used for both macro / micro-hardness measurement to evaluate specific microstructural features like martensite or bainite, as well as checking the effectiveness of heat treatment and case hardening processes. As per materials (i.e. in light of guessed hardness), normal load is applied. The indenter penetrates the surface. The two diagonals square indentation are measured. Average diagonal length of indentation for a given major load (i.e. indirectly) indicates the hardness (Fig. 2.28). In general, a larger diagonal length of indentation shows low hardness and the vice versa.

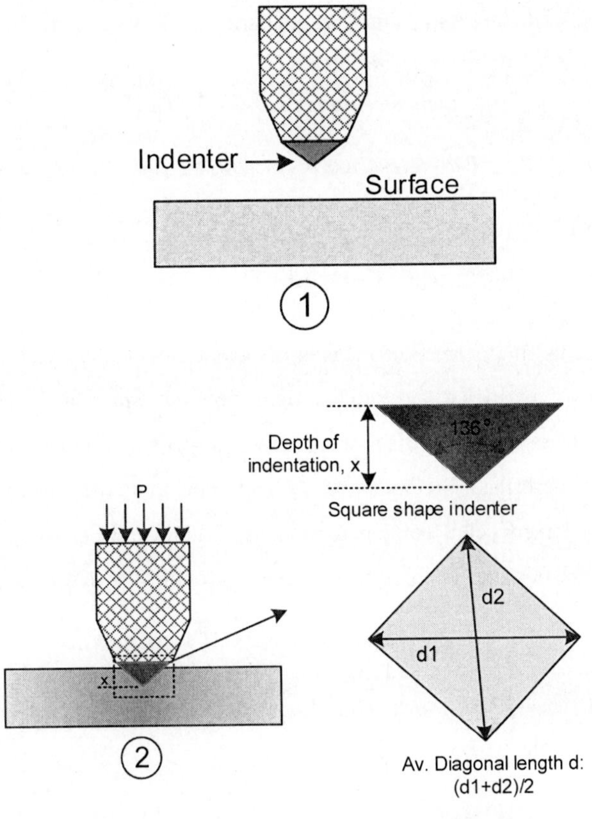

Figure 2.28: Schematic diagram sequential steps of Vicker's hardness test 1) prepare surface ready for testing, 2) normal load applied through indenter resulting in desired indentation showing closure view of indentation and different dimensions

Considering the strain-hardening coefficient, relationship between ultimate tensile strength and Vickers hardness is given below.

$$\textbf{\textit{Ultimate tensile strength}} = \left(\frac{H}{3}\right)\left(\frac{n}{0.217}\right)^{n}$$

where n is strain-hardening coefficient, H is the Vickers hardness, and UTS is the ultimate tensile strength.

Yield strength (YS), Vickers hardness *(H)*, and strain-hardening coefficient (*n*) relationship can be used to predict the yield strength of aluminium, copper and steel.

$$\textit{Yield strength}=\left(\frac{H}{3}\right)\left(\frac{1}{10}\right)^{n}$$

Additionally, the following linear equations have also be proposed to calculate the ultimate tensile strength and yield strength using Vickers hardness. Where yield/ultimate strength is in MPa and Vickers Hardness (H) is in kg/mm^2.

$$\textit{Yield strength} = -90.7 + 2.876\,H$$

$$\textit{Ultimate strength} = -99.8 + 3.734\,H$$

UNIT SUMMARY

The mechanical properties of material plays an important role in efficient manufacturing and design of components. Importance of mechanical properties namely yield strength, ultimate strength, modulus of elasticity and rigidity, ductility, toughness and hardness have been presented using suitable schematic diagrams. Additionally, using fundamentals the generalized hooks law has been described. Methods to examine the behaviour of materials under different load conditions e.g. tension, compression, torsion, impact load have also been covered. Further, derivative characteristics like resilience, toughness, strength vs. hardness relationships have also be included.

EXERCISE

Multiple Choice Questions

Questions for self-assessment

1. The mechanical property of the material is
   a. Ductility
   b. Atomic packing factor
   c. Microstructure
   d. Melting temperature
2. Toughness of a martial obtained from the tensile test depends on the combination of
   a. Hardness and tensile strength
   b. Hardness and ductility
   c. Tensile strength and ductility
   d. Modulus of elasticity and tensile strength
3. Necking in a ductile metal during the tensile test begins at
   a. Elastic limit
   b. Yield point
   c. Ultimate point
   d. Fracture point
4. Impact toughness of a materials is evaluated by
   a. Sharpy test
   b. Izod test
   c. Hammer test
   d. All of these
5. There are four metals A, B, C and D show 20, 40, 60 and 100 J impact energy absorbed during toughness. The toughest metal is
   a. A
   b. B
   c. C
   d. D
6. The type of stress induced in a materials under torsional load will be
   a. Bending

b. Tensile

c. Compressive

d. Shear

7. A cylindrical shaft is subjected to a torque of 200 Nm. The maximum shear stress will be at

a. Centre axis

b. Surface

c. Between centre axis and surface

d. Same at the all locations

8. A plateau in load vs deformation plot is observed during the compression test of

a. Porous metal

b. Metallic foam

c. Cardboard

d. All of these

9. Application of a minor load is a part of procedure during hardness test by

a. Brinnel method

b. Rockwell method

c. Knoop method

d. Vickers method

10. Square shape indentation is observed in case of

a. Brinnel method

b. Rockwell method

c. Vickers method

d. All of these

Answers of Multiple Choice Questions

Key for MCQ: 1 a, 2 c, 3 c, 4 b, 5 d, 6 d, 7 b, 8 d, 9 b, 10 c

Short and Long Answer Type Questions

1. What is the significance of ductility and yield strength?
2. Enlist the properties that can be obtained from a tensile test.
3. What is the meaning of lateral strain in tensile test sample?
4. Why does the true stress always greater than engineering stress?
5. Compare the Hook's law and generalised Hook's law.
6. How to determine the resilience from a tensile of material.

7. What is off-set stress and when it is used to establish the yield point?
8. What are factors affecting the shear strain in component subjected to torsional load?
9. What is the significance of hardness measurement of a material?
10. What inferences can be drawn from the "low impact energy absorbed" by sample during toughness test?
11. Compare the different hardness test methods of various technical aspects.
12. What is significance of minor load in Rockwell hardness test?
13. Elaborate the hardness and tensile strength relationship.
14. Describe the principle of any Impact toughness test method.
15. Compare the stress vs strain curves for metals and cardboard under compressive loading.

Numerical Problems

1. The tensile test conducted on a cylindrical sample of 10 mm diameter and 50 mm gauge length. During the test, sample showed commencement of plastic deformation at 0.5 mm elongation and corresponding tensile load was 40 kN. The maximum load taken by the sample was 70 kN which in turn return resulted in necking. The sample at fracture showed final (gauge) length of 70 mm in abutting position and fractured end diameter of 6 mm. Assume poisons ratio 0.4, calculate the following:
    - Yield strength
    - Ultimate strength
    - % elongation
    - % reduction
    - Strain at the yield point
    - True stress at the time fracture
    - True strain at the time fracture
2. The tensile test conducted on a rectangular section sample of size 10 mm thick and 25 width with 100 mm gauge length made of metal having modulus of elasticity 2 X 10^6 N/mm^2. Assume poisons ratio 0.4, calculate the engineering and true strain at tensile load 20 kN.
3. The compression test conducted on a cylindrical sample of 30 mm diameter and 50 mm gauge length. The sample showed reduction in length of 5 mm at 20 kN tensile load. The sample length at crushed fracture stage is 23 mm with maximum load of 38 kN.

Assume poisons ratio 0.4, calculate the engineering and true strain at 20 kN load and fracture stage.

4. The torsion test conducted on a cylindrical sample of 10 mm diameter and 50 mm gauge length. The sample fixed at one end while other end was subjected to a torsional load. The sample during torsion showed angle of twist 5° at 700 Nm torque. Calculate the modulus of rigidity of material.
5. A cuboid of side 10 mm subjected tri-axial tensile load of 100 kN, 60 kN and 40 kN. Assume poisons ratio 0.3, calculate the strain in the direction of maximum and minimum load.
6. The tensile test showed stress at elastic limit 160 MPa for the sample material is 2 X 10^6 N/mm^2. Calculate the resilience of the sample material.

PRACTICAL

1. Conduct tensile test of mild steel. Plot engineering stress vs strain curve and obtain elastic modulus, elastic limit, yield strength, ultimate strength, ductility in terms of % elongation and % reduction in area, resilience and toughness.
2. Conduct compression test on Aluminium / Cu of 20 mm diameter and 20 mm long cylindrical block. Plot engineering stress vs strain curve and obtain elastic modulus, elastic limit, "2% reduction" yield point.
3. Conduct torsion test on mild steel of 20 mm diameter and 100 mm long cylindrical block. Plot torque vs angle of twist (shear stress vs strain) curve and obtain modulus of rigidity, elastic limit, yield and ultimate point, and resilience.
4. Conduct Izod impact test on standard sample of mild steel and cast iron. Compare the impact energy absorbed, fracture surface morphology. Include your comments with regard to expected performance of two under impact load conditions.
5. Conduct any hardness test (Brinnel, Vickers) on steel and cast iron. Estimate tensile strength of two metals. Compare the hardness and corresponding indentation morphology. Include your comments with regard to expected performance of two under impact and abrasive wear conditions.

KNOW MORE

Explore the historical accidents like Titanic and Liberty ships, which primarily caused by unsuitable combination of mechanical properties for service conditions (sub-zero temperature and impact). There have been many historical failures (e.g. nuclear reactors, pressure vessels, aircraft, spacecraft, boilers etc.) occurred due to inappropriate choice of materials and mechanical properties. Learning from the failure helps us to avoid reoccurrence of the similar failures in future.

REFERENCES AND SUGGESTED READINGS

Suggested resources for further reading/learning

1. Callister, Materials science and engineering, Wiley, Publication (2014)
2. S D Avner, Introduction to Physical Metallurgy, McGraw Hill, Publication (2002)
3. GE Dieter, Mechanical metallurgy, McGraw Hill, Publication (1989)
4. D K Dwivedi, Fundamentals of Metal Joining, Springer nature, Singapore (2022)
5. D K Dwivedi, Surface Engineering, Springer Nature, Singapore (2018)
6. Ratna Kumar, NPTEL Course "Basics of Materials Engineering" https://onlinecourses.nptel.ac.in/noc20_me78/preview

Dynamic QR Code for the Unit 2

Unit 3

Theories of Failures under Static Loading and Fatigue Behaviour

Ductile and brittle failure, fracture toughness, fatigue, fatigue life, mean stress & fatigue

Unit Specific / Learning Objective

The objective of this unit is to talk about the following aspects

- To identify the state of failure of a component
- To learn the different ways by which a mechanical component can fail
- To introduce the concept of the fracture mechanics
- To learn classically and modified Griffith's theory of brittle fracture
- To understand the importance of the various theories of failure
- To learn how to distinguish the mode of a failure
- To understand the significance of the various failure theory and their applicability
- To introduce the concept of fatigue strength and the fatigue life
- To learn about the different stages of the fatigue fracture

Additionally, a few fundamental questions for self-assessment based on fundamentals are included in this chapter in the form of recall, application, comprehension, analysis, and synthesis. There are further suggested readings and references for deep learners and reader's assistance.

Rationale

The failure of material can occur in different stages. The failures occurring in the manufacturing stage adversely affect productivity due to the rejections and the related rework cost. The failures during the service affect availability, production, inconvenience, and cost. Further, the failures of the critical component can lead to the loss of life and property. Therefore, it is important to understand the stress conditions under which mechanical component failure can occur. The failure theory helps in designing the suitable component of material as per our service conditions to avoid failures. Therefore, it is important to understand various failure theories for static loading developed for ductile and brittle materials. The components in real life are subjected to fluctuating loads causing fatigue and premature failure. Therefore, it is important to have a systematic understanding of fatigue loading and its characteristics, stages of fatigue fracture, fatigue strength, and fatigue life to design the components so as to avoid failures under fluctuating load conditions. At the manufacturing stage, it is desired to evaluate the soundness of the component so that it can be processed further to make a product. A similar examination of the mechanical component is also needed during the service to examine the presence and growth of cracks, if any, to avoid a catastrophic fracture and take suitable corrective measures on time. Non-destructive testing helps to examine the component during both manufacturing and service. Therefore, it is important to learn about common non-destructive testing techniques.

Pre-Requisites

A course on Physics / Mechanics: stress, strain (Class XII)

Learning outcomes

U3-O1: Ability to choose design criteria for developing the failsafe design of ductile and brittle materials using the appropriate theory of failures

U3-O2: Ability to identify stress conditions that can lead to the failure of a component

U3-O3: Ability to determine the critical crack size using fracture toughness, which can lead to catastrophic failure

U3-O4: Ability to determine the fatigue strength and fatigue life of a component using the S-N curve and estimate the allowable for taking strength in terms of the stress amplitude using mean stress of fatigue loading and tensile properties

U3-O5: Ability to choose the suitable non-destructive technique to evaluate the surface and subsurface and apply the non-destructive techniques to establish the size, shape and location of the defects

Unit-1 Outcomes	***EXPECTED MAPPING WITH COURSE OUTCOMES*** *(1- Weak Correlation; 2- Medium correlation; 3- Strong Correlation)*					
	CO-1	**CO-2**	**CO-3**	**CO-4**	**CO-5**	**CO-6**
U3-O1	*1*	2	3			
U3-O2	*1*	2	3			
U3-O3	*1*	2	3			
U3-O4	*1*	2	3			
U3-O5	*1*	*2*	3			

Course Objective

1. Students will be able to identify crystal structures of various materials, and one could understand the defects in such structures.
2. Understand how to tailor the material properties of ferrous and non-ferrous alloys?
3. How to quantify mechanical integrity and failure in materials?

3.1 Introduction

The mechanical components/products are designed to work under certain service conditions to give satisfactory performance for a designed life. These service conditions include temperature, load, environment, etc., varying significantly depending upon the product type and its application. A perfectly designed and manufactured product offers useful life as per the designed life. However, as per service conditions, a component experiences gradual degradation during service in the form of wear and tear, deterioration in microstructure and mechanical properties, loss of dimension like thinning, elastic and deformation cracking and even fracture.

3.1.1 Failure of a component

A component is considered to have failed in three different situations a) the component does not work at all, b) a component works but does not give the intended performance and c) the working of the component is unreliable, erratic and risky to life and property.

From the mechanical point of view, failures of a component can occur under any of the following three conditions when the component experiences a) elastic deformation beyond the acceptable limits, b) plastic deformation beyond the acceptable limits leading to an undesirable change in dimensions and geometry of the product, and c) fracture due to fatigue, creep, overloading etc. in two more parts.

3.1.2 Failure under different loading conditions

The mechanical components are designed either based on static or dynamic loading while taking other service conditions into account. The static loading can be tensile, compressive, and torsional, wherein the magnitude and direction of the load largely remain constant. The dynamic loading, on the other hand, can be in the form of fatigue, impact loading

The design of components subjected to uniaxial loading (e.g. only on tensile/compressive/torsional load) is somewhat easier as it is based on the consideration of permissible stress (for a given type of stress as per loading) to determine the load resisting cross-sectional area of the component. The choice of permissible mechanical property (yield and ultimate strength, ductility, toughness) however depends on the material, and factor of safety as per the criticality of the application of the component to be designed (Table 1). For example, the design of components made of ductile materials (> 5% elongation) is based on yield strength, while the components of brittle materials are designed based on ultimate strength (Table 2). Further, the component wherein even elastic deformation leads to failure of the component is designed on the basis of modulus of elasticity (E) or rigidity (G) as per loading conditions.

Table 3.1: Type loading and design criteria

Type of loading		Mode of Failure		
Nature of load	Stress	Elastic deformation	Plastic deformation	Fracture
Tensile	Tensile	Modulus of elasticity	Yield strength	Ultimate strength
Compressive	Compressive	Modulus of elasticity	Yield strength	Ultimate strength
Torsion	Shear	Modulus of rigidity	Yield strength	Ultimate strength

Table 3.2: Type of material and design criteria

Type of materials	Design criterion	Failure mode
Yield	Yield/shear strength	Plastic deformation
Brittle materials	Ultimate strength	Fracture
High strength and tough materials for critical applications	Fracture toughness	Mixed mode

3.2 Ductile and brittle failure

The components designed based on static loading and considering a suitable type of allowable stress can fail in a ductile or brittle manner depending upon the type of material and service conditions exposed. The categorization of ductile or brittle failure is generally done based on the extent of plastic deformation prior to the fracture. However, it can also be done based on the micro-mechanism of fractures like dimples, cleavage facets, inter-granular fractures, and striations. Ductile materials show dimple fracture, while brittle materials exhibit cleavage facets intergranular fracture. The coverage of the failures on the basis of microscopic mechanisms of fracture is beyond the scope of this book. Real-life service conditions to which a component is exposed can vary significantly—in terms of loading on the component. If load crosses the upper limit accidentally and inadvertently, then "overload fracture in the form of ductile or brittle failure occurs.

High ductility material (>5% elongation) under static load conditions usually fails in a ductile manner following the significantly plastic deformation. The ductile fracture begins with the nucleation of micro-voids somewhere in the middle of a cross-section at the necking stage. These voids continuously grow with the increase of load, and eventually, the coalescence of these voids results in a significant reduction in load resisting cross-sectional area leading to overload ductile fracture with a great reduction in cross-sectional area (Fig. 3.1).

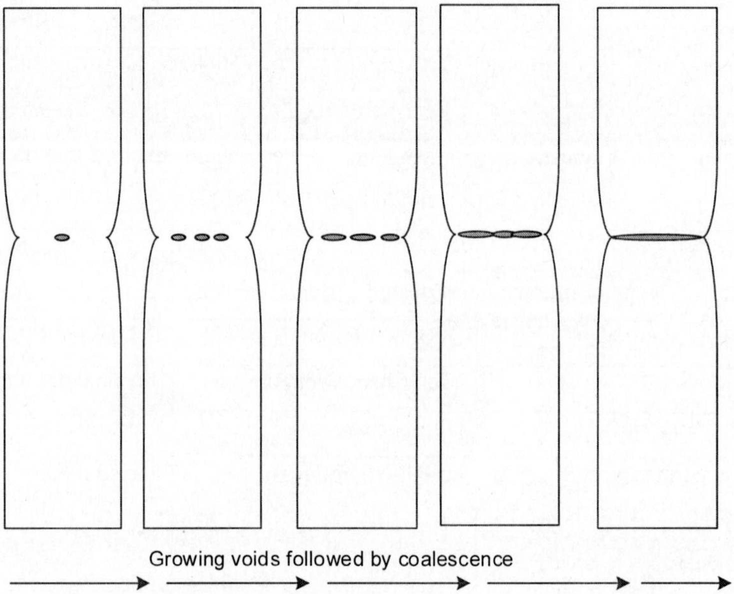

Figure 3.1 Schematic diagram shows ductile fracture mechanism through sequential micro-void nucleation, growth, coalescence, and fracture.

However, under certain unfavourable service conditions like very high rate of loading, low-temperature exposure (below ductile to brittle transition temperature), high-temperature exposure leading to the embrittlement of materials due to unexpected and undesirable metallurgical transformation can result in brittle failure of ductile materials. Therefore, for ease of understanding, all the failures involving significant yielding can be considered ductile failures. The ductile failure gives enough signals and indication before fracture, while the brittle fracture is catastrophic and almost within no time and without any significant indication of plastic deformation.

The ductile or brittle failures can be grouped based on a) extent of elongation (%) or reduction in area (%) of the fracture components and b) fracture surface morphology. A low % age elongation and reduction in area (<5%) indicates brittle failure. However, this is indicative and is not a very hard and fast demarcating value (i.e. 5%) to categories modes of failure (Fig. 3.2). Real-life service conditions for which a component is exposed can vary significantly. Loading on the component if crosses the upper limit of loading accidentally and inadvertently then overload fracture in the form of ductile or brittle failure can occur.

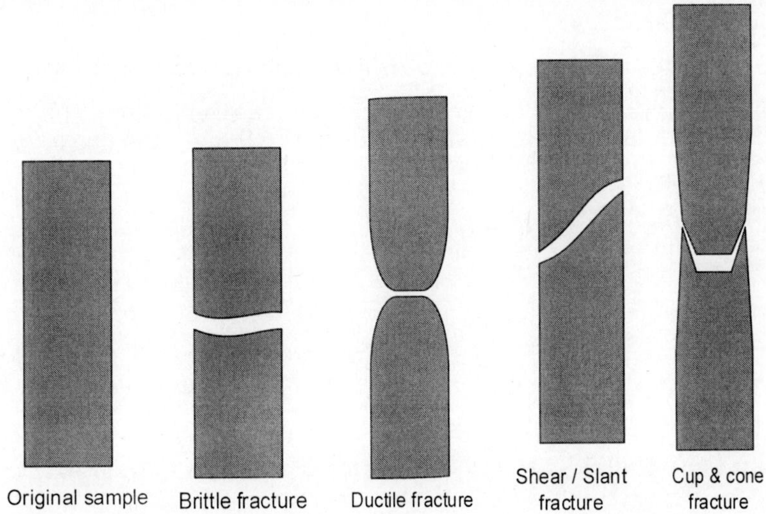

Figure 3.2 Schematic diagram of different fracture modes a) original sample, b) brittle fracture, c) ductile fracture-excessive necking, d) ductile fracture- slant fracture, and e) ductile fracture- cup-cone fracture.

In the case of ductile failures, when the plastic deformation of a component goes beyond the acceptable value, the component is considered to have failed. In the case of brittle failure, the fracture of the component into two or more is considered as a failure.

3.3 Need of Theories of Failure

A component subjected to uniaxial loading successfully designed based on allowable given type of stress establishes the load resisting cross-sectional area. However, a component subjected to biaxial or multi-axial loading, either single or different types of load (tensile, compression, torsion) makes the successful design a little complicated because of the varying response of material (as compared to uniaxial loading) in terms of the modulus of elasticity/rigidity, yield and ultimate strength, % age elongation and reduction in area, impact toughness under such loading conditions. Under such conditions, stresses leading to unacceptable elastic/plastic deformation and fracture differ from allowable uniaxial yield strength, ultimate strength, etc., obtained through uniaxial tensile, compression and torsion tests. In general, materials under biaxial or multi-axial loading conditions tend to behave more like brittle material as compared to uniaxial loading. The stress at which the yield of materials begins (yield point) varies with loading conditions as evident from the following. This behaviour of materials under multi-axial loading makes it difficult to find an appropriate criterion for the design of the component to avoid any premature failure during the service.

3.4 Principal stress and Mohr circle

The figure shows the stress element in plane strain conditions (2D component) at a point in a mechanical component. It can be observed that there are two normal stresses (σ_x, σ_y) and shear stress (τ_{xy}) on the faces of the element in consideration (Fig. 3.3).

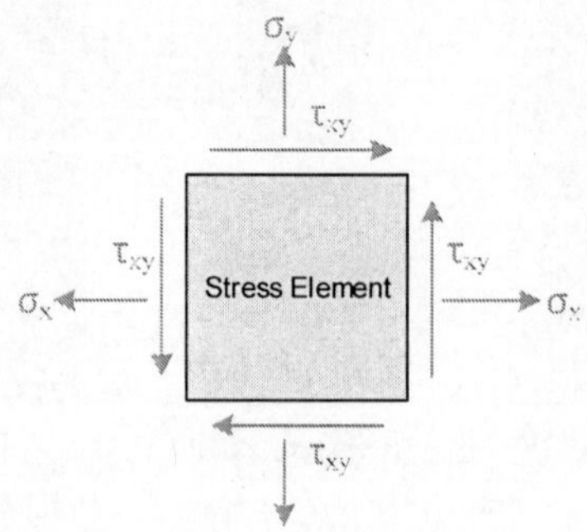

Figure 3.3 Schematic diagram shows two normal tensile stress and shear stress on a stress element.

If the element is in the line of the loading axis, Then there would be only normal stresses (sigma x) while normal stress sigma y and shear stresses will be zero (Fig. 3.4).

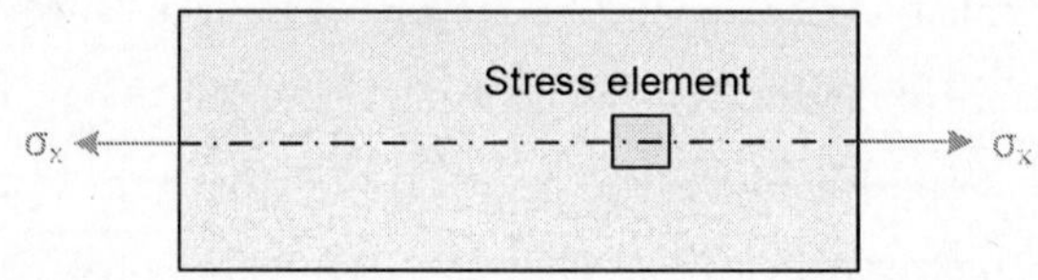

Figure 3.4 Schematic diagram shows that stress element on the axis of unidirectional stress (σ_x) with zero shear stress.

If the stress element is rotated in an anticlockwise direction, then the normal stress and shear stress continuously change (Fig. 3.5). Careful observation of the variation in the stress as a function of the angle of rotation theta indicates that the normal stress is maximum when shear stress is zero and vice versa (Fig. 3.6). Further, the maximum and minimum normal stresses are separated by 90 degrees. If one face of the stress element has the maximum stress, then the adjacent face at 90 degrees will have the minimum principal stress. Shear stress is zero at the location where maximum and minimum normal stresses occur. The plane of such angles on which the stress element has maximum and minimum normal stress are called the principal planes, and the corresponding stresses are called maximum and minimum principal stress. The rotation of an element by 360 degrees will result in the same stress state we had in the beginning in the element. It is important to note that the actual stress state of the component as a whole remains the same and the above variation just represents how the stress-state of an element is affected if it is rotated by different angles.

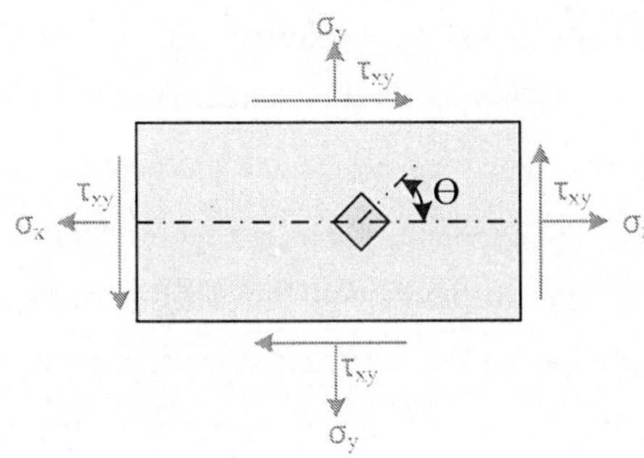

Figure 3.5 Schematic diagram shows two normal tensile stress and shear stress on stress element rotated anticlockwise by angle Θ.

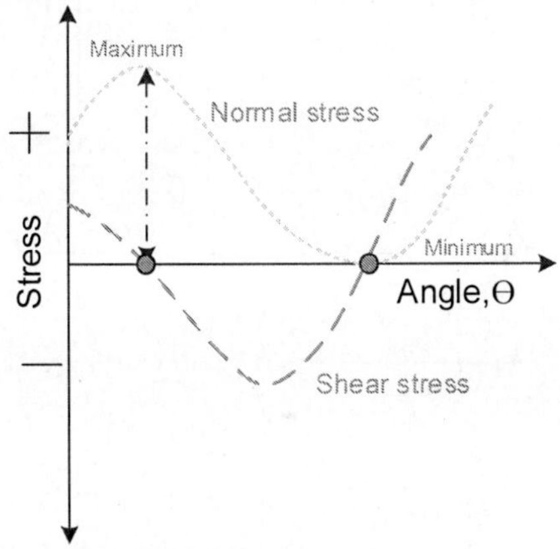

Figure 3.6 Schematic diagram shows variation in normal stress and shear stress on stress element as a function of rotation angle Θ in anticlockwise direction.

$$\sigma = \frac{\sigma x + \sigma y}{2} + \frac{\sigma x - \sigma y}{2}\cos 2\theta + \tau xy \sin 2\theta$$

$$\tau = -\frac{\sigma x - \sigma y}{2}\sin 2\theta + \tau xy \cos 2\theta$$

The principal stresses are very important, and many failure theories have been developed based on the principal stresses. The principal stresses are the maximum and minimum normal stresses at a particular point where the shear stress is zero. The maximum and minimum principal stresses are denoted by σ_1 and σ_2 respectively. The maximum principal stress (σ_1) helps in predicting the failures, and the plane of the maximum principal stress indicates the potential plane where failures can occur.

Mohr Circle

Mohr circle is a graphical method to determine normal and shear stresses at different orientations without stress transformation equations (Fig. 3.7). The construction of the Mohr circle involves drawing two lines, one horizontal and another vertical. The horizontal line represents normal stresses, and the shear stresses are shown by the vertical line. Positive and negative shear stresses are shown below and above the horizontal line, respectively. Shear stress is considered positive if it tends to rotate the stress element in the anticlockwise direction, and it is considered negative if it tends to rotate the stress element in the clockwise direction.

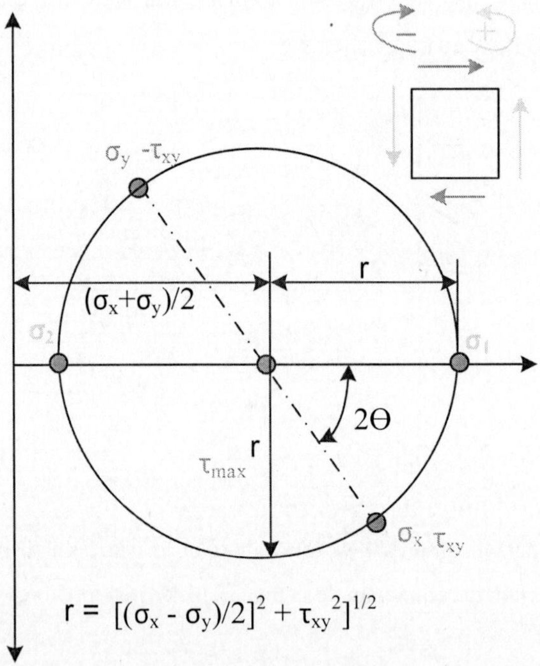

Figure 3.7 Schematic diagram shows Mohr circle with normal stress, shear stress and principal stress, along with various parameters of the circle and their significance.

Every point on the Mohr circle indicates specific information. The radius of the circle indicates the maximum shear stress, which can be, measured directly using a suitable scale and the same also can be exactly obtained using the following equation. The maximum and minimum principal stress can also be obtained from the more circle by finding out the points where the circle crosses the horizontal line where shear stresses are zero.

Max. shear stress (r): $[(\sigma_x - \sigma_y)/2]^2 + \tau_{xy}^2]^{1/2}$

Maximum and minimum principal stress (σ_1, σ_2): centre of the circle $\pm$ radius of the circle

$$(\sigma_1, \sigma_2) = (\sigma_x + \sigma_y)/2 \pm [(\sigma_x - \sigma_y)/2]^2 + \tau_{xy}^2]^{1/2}$$

In the three-dimensional stress conditions of an element, there will be three principal stresses (s1,s2, s3). These principal stresses are in decreasing order from maximum (σ_1) to minimum (σ_3). Mohr circle for three-dimensional stress state is obtained by drawing three different circles, as shown in Fig. 3.8. Various normal stresses and shear stresses are found in the sectioned zone of the Mohr circle.

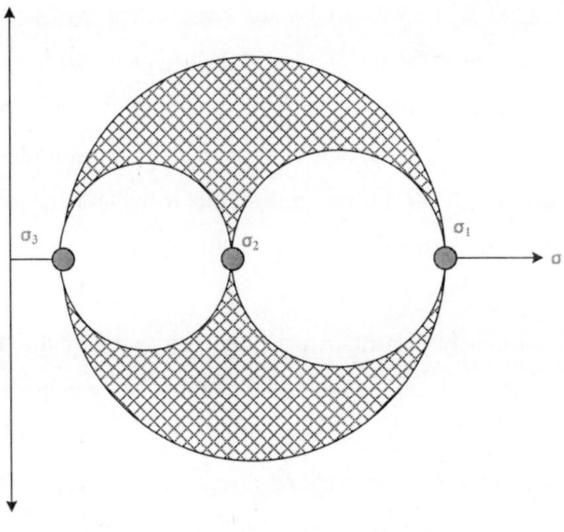

Figure 3.8 Schematic diagram shows Mohr circle with three normal stresses on a stress element and their significance.

3.5 Theories of Failure

Therefore, many theories of failure have been proposed for ductile and brittle materials; which form a basis for the design of the component subjected to biaxial or multi-axial loading condition. The most common theories include Maximum Shear Stress (Tresca) theory, Maximum principle stress (Rankine's), Maximum distortion energy (Von-mises), Maximum & total strain energy, Mohr-coulomb and modified Mohr-coulomb theory. The fail-safe stress under combined loading conditions for different types of material (ductile and brittle) can be obtained using relevant theories of failure.

The theories of failure help to predict the stress conditions when failure (fracture/yielding as per material and loading conditions) will occur in terms of easily measurable mechanical properties yield and ultimate strength obtained from the tensile test.

A single theory does not suit to explain failure under all types of loading and materials and designing the component. For brittle materials, the Maximum principle stress (Rankine's), Mohr-coulomb and Modified Mohr-coulomb theory are found used for the design of component under biaxial and tri-axial loading, while the Maximum Shear Stress (Tresca), Maximum distortion energy (Von-mises), Maximum & total strain energy theory are found suitable for ductile materials.

In the case of brittle materials, the failure occurs by a fracture due to overloading beyond the ultimate strength (S_{ut}) of material under tension, compression and torsion while the ductile materials fail by yielding or plastic deformation. Since the plastic deformation occurs due to slip and twining mechanisms wherein "shearing" plays a crucial role in causing the slip of one layer of atoms over another on the slip plane to cause the plastic deformation, therefore, theories of failures of ductile materials consider shear strength (as a criterion for yielding) in terms of yield strength (S_{yt}).

These theories assume that material is a) free from dis-continuities like cracks, and b) isotropic and homogeneous. Further, these theories consider the principal stresses only induced due to uniaxial biaxial and multi-axial loading at a point in 3-dimensional space, which makes the location point of interest irrelevant. From the design and utility point of view, the outcome of these theories is expressed in yield strength and ultimate strength of the metal obtained through tensile, compression and torsion test.

Simplest form of a theory of failure is as under

$$f(\sigma_1, \sigma_2, \sigma_3) = S_{yt}, S_{ut}$$

3.5.1 Maximum Principal Stress Theory (Rankine's Theory)

According to the maximum principal stress theory, failure occurs when the maximum or minimum principal stress exceeds the yield or ultimate strength (S_{yt}, S_{ut}) of metal obtained from a simple tensile test of the material. This theory is applied to brittle material under all biaxial and tri-axial loading conditions. However, this theory does not work effectively in developing the fail-safe design of ductile materials because ductile materials are weak in shear.

$$(\sigma_1) > S_{yt}, S_{ut}$$

$$(\sigma_3) > -S_{yt}, -S_{ut}$$

If σ_1 is positive (under tension) then yield or ultimate strength of materials is taken from tension test (S_{yt} or S_{ut}) and when σ_1 is negative (under compression) then yield or ultimate strength of materials in form of compression (S_{yc} or S_{uc}).

To develop the fail-safe design, if we take a factor of safety (F) to design of component is N then the maximum principal stress (σ_1) should be $\leq$ Permissible stress ($\sigma_{allowable}$)

Permissible stress ($\sigma_{allowable}$) = Failure stress/factor of safety = $\frac{S_{yt}}{F}$ or $\frac{S_{ut}}{F}$

$$\textbf{Maximum principal stress } (\sigma_1) < \frac{S_{yt}}{F} \text{ or } \frac{S_{ut}}{F}$$

To develop the fail-safe design of brittle material using principal stress theory, the yield and ultimate strength of material obtained from the tensile is considered because brittle materials are found to be weaker under tension than compression (Fig. 3.9).

Moreover, this theory works even for ductile material under the following stress state conditions a) the maximum shear stress ($\tau_{max} = \sigma_1 \div 2$) is half of the maximum principal stress under uniaxial loading conditions, b) biaxial stress state when both principal stresses are of the same type (absolute $\tau_{max} = \sigma_1 \div 2$) and c) hydrostatic stress condition wherein shear stress in all the planes is found be zero.

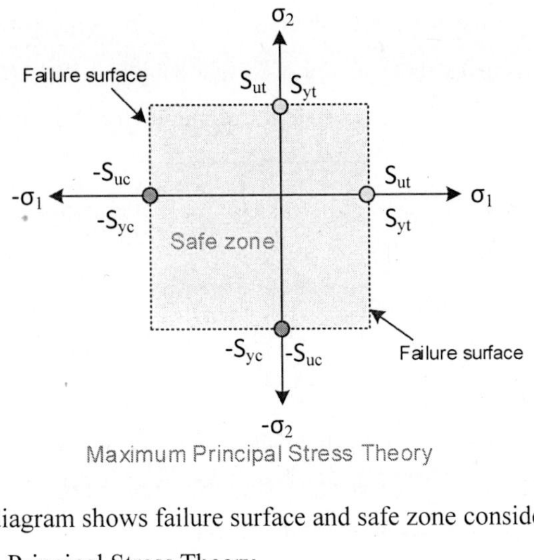

Figure 3.9 Schematic diagram shows failure surface and safe zone considering two normal stress as per Maximum Principal Stress Theory.

3.5.2 Maximum Shear Stress (Tresca) Theory

As per the maximum shear stress theory, the failure (by plastic deformation or yielding) of a mechanical component made of ductile material (subjected to uniaxial, biaxial or tri-axial loading) occurs when the maximum shear stress at a point in the material exceeds the value equivalent to the maximum shear stress developed in the standard specimen during the tensile test. As we know that the specimen during the tensile test is subjected to a uniaxial condition of stress (σ_2 & $\sigma_3 = 0$) while maximum principal stress ($\sigma_1 = S_{yt}$).

As a result, the maximum shear equals 50% of the difference between both the maximum and minimum principal stresses. Therefore, the maximum shear stress in a simple tensile test specimen is half of the yield strength of metal in a tensile test. Therefore, the failure condition according to maximum shear stress theory is:

$$\text{Absolute } T_{max} = (S_{ys})_{TT} \text{ or } = S_{yt}/2$$

Where S_{ys} is yield strength in shear under tri-axial stress-state, which in unknown and S_{yt} is the yield strength in tension test.

A Hexagon graphically illustrates stress distribution, indicating that the material will attain its elastic limit once the stresses (σ_1 and σ_2) exceed this area. Further, the maximum distortion-strain energy approach evaluates the yielding condition better than the Tresca theory.

Step 1: Determine the three principal stresses (σ_1, σ_2, and σ_3) from the tri-axial stress system using principal stress equations or Mohr's circle method.

Step 2: Find out the maximum (σ_1) and the minimum (σ_3) principal stresses

Step 3: Determine the value of the maximum shear stress $\tau_{max} = (\sigma_1 - \sigma_3)/2$.

Step 4: Find out the allowable stress value of the material; allowable stress= σsy /N or σy /2N as mentioned above (N=Factor of safety)

Step 5: Compare the value calculated in step 3 with the allowable value found in step 4. If the Value at step 3 is less than the allowable value at step 4, then the design is safe as per the maximum shear stress theory.

For the fail-safe design, the maximum shear stress $(\tau_{max.})_{TT}$ induced at a point in a metal during the tensile should be less than allowable shear stress ($\tau_{allowable}$) while considering factor of safety (F) as shown in Fig. 3.10.

$\tau_{allowable} = (\tau_{max.})_{TT} / F = (S_{ys})_{TT} / F = S_{yt} / 2.F$

Absolute $\tau_{max.} \leq (S_{ys})_{TT} / F = S_{yt} / 2.F$

Uniaxial stress state

Absolute $\tau_{max.} \leq S_{yt}/2.F$

Biaxial stress state (σ_3=0)

Higher of [absolute value of $\sigma_1/2$, $(\sigma_1-\sigma_2)/2$] $\leq S_{yt}/2.F$

Absolute value of $\sigma_1 \leq S_{yt}/F$ when σ_1 and σ_2 are similar in nature

Absolute value of $(\sigma_1-\sigma_2)$] $\leq S_{yt}/F$ or when σ_1 and σ_2 are different in nature

Tri-axial stress state

Higher of [absolute value of $(\sigma_1-\sigma_2)/2$, $(\sigma_2-\sigma_3)/2$, $(\sigma_3-\sigma_1)/2$] $\leq S_{yt}/2.F$ or

Higher of [absolute value of $(\sigma_1-\sigma_2)$, $(\sigma_2-\sigma_3)$, $(\sigma_3-\sigma_1)$] $\leq S_{yt}/F$

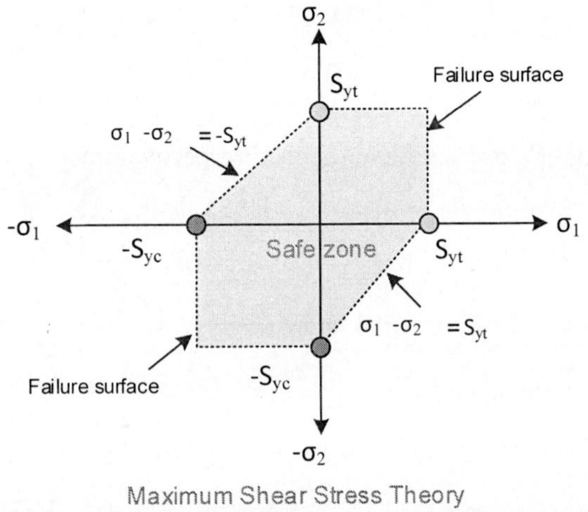

Figure 3.10 Schematic diagram shows failure surface and safe zone considering two normal stress as per Maximum Shear Stress Theory.

3.5.3 Maximum Distortion Energy Theory (Von Mises)

As per the maximum distortion energy theory, the failure (by plastic deformation or yielding) of a mechanical component made of ductile material (subjected to uniaxial, biaxial or tri-axial loading) occurs when the maximum distortion energy per unit/volume is greater than distortion energy per unit volume at the yield point during the tensile test. As we know that the specimen during the tensile test is subjected to a uniaxial condition of stress (σ_2 & σ_3 = 0) while maximum principal stress ($\sigma_1 = S_{yt}$). For fail-safe design, therefore, the maximum distortion energy per unit/volume should be equal to or less than the distortion energy per unit volume at the yield-point during the tensile test (Fig. 3.11).

Total strain energy / unit = (Volumetric strain energy / volume) + (Distortion energy / volume)

Distortion energy / volume = (Total strain energy / unit) - (Volumetric strain energy / volume)

Total strain energy / volume (within elastic limit): area under elastic curve: $\frac{1}{2}\, stress\ (\sigma) strain\ (\varepsilon)$

For tri-axial stress state

Total strain energy / volume (within elastic limit): $\frac{1}{2}\sigma_1.\varepsilon_1 + \frac{1}{2}\sigma_2.\varepsilon_2 + \frac{1}{2}\sigma_3.\varepsilon_3$

Since $\varepsilon_1 = \frac{1}{E}[\sigma_1 - \mu(\sigma_2 + \sigma_3)]$,

$\varepsilon_2 = \frac{1}{E}[\sigma_2 - \mu(\sigma_1 + \sigma_3)]$,

$\varepsilon_3 = \frac{1}{E}[\sigma_3 - \mu(\sigma_1 + \sigma_2)]$

Therefore, total strain energy / volume (within elastic limit): $\frac{1}{2}\sigma_1.\varepsilon_1 + \frac{1}{2}\sigma_2.\varepsilon_2 + \frac{1}{2}\sigma_3.\varepsilon_3$ can be written as under by substituting values respective strains

Total strain energy / volume (within elastic limit)

$= \frac{1}{2}\sigma_1.\frac{1}{E}[\sigma_1 - \mu(\sigma_2 + \sigma_3)], + \frac{1}{2}\sigma_2.\frac{1}{E}[\sigma_2 - \mu(\sigma_1 + \sigma_3)], + \frac{1}{2}\sigma_3.\frac{1}{E}[\sigma_3 - \mu(\sigma_1 + \sigma_2)]$

$= \frac{1}{2E}[\sigma_1^2 + \sigma_2^2 + \sigma_3^2 - 2\mu(\sigma_1.\sigma_2 + \sigma_2.\sigma_3\ \sigma_3.\sigma_1)]$

Volumetric strain energy per unit volume = $\frac{1}{2}$ Average stress X volume strain

$= \frac{1}{2}[(\sigma_1 + \sigma_2 + \sigma_3)/3]\ X\ [\frac{1-2\mu}{E}(\sigma_1 + \sigma_2 + \sigma_3)] = [\frac{1-2\mu}{6E}(\sigma_1 + \sigma_2 + \sigma_3)^2]$

Distortion energy / volume = (Total strain energy / unit) - (Volumetric strain energy / volume)

$= \frac{1}{2E}[\sigma_1^2 + \sigma_2^2 + \sigma_3^2 - 2\mu(\sigma_1.\sigma_2 + \sigma_2.\sigma_3\ \sigma_3.\sigma_1)] - [\frac{1-2\mu}{6E}(\sigma_1 + \sigma_2 + \sigma_3)^2]$

$= [\frac{1+\mu}{6E}[(\sigma_1 - \sigma_2)^2 + (\sigma_2 - \sigma_3)^2 + (\sigma_3 - \sigma_1)^2]$

Distortion energy per unit volume in metal at yield point during tensile test can be obtained using $\sigma_1 = S_{yt}/F$, $\sigma_2 = \sigma_3 = 0$

Distortion energy per unit volume at the yield point during tensile test= $\frac{1+\mu}{3E}(\sigma_1)^2 == [\frac{1+\mu}{3E}(S_{yt}/F)^2]$

For fail-safe design, the distortion energy/volume under tri-axial stress state $\leq$ Distortion energy per unit volume at yield point during tensile test

$[\frac{1+\mu}{6E}[(\sigma_1 - \sigma_2)^2 + (\sigma_2 - \sigma_3)^2 + (\sigma_3 - \sigma_1)^2] \leq [\frac{1+\mu}{3E}(S_{yt}/F)^2]$

For fail-safe design, the distortion energy / volume under biaxial stress state (σ_3=0) ≤ Distortion energy per unit volume at yield point during tensile test

$$\left[\frac{1+\mu}{6E}\left[(\sigma_1-\sigma_2)^2+(\sigma_2-\sigma_3)^2+(\sigma_3-\sigma_1)^2\right]\right] \leq \left[\frac{1+\mu}{3E}(S_{yt}/F)^2\right]$$

Conversely, $[\sigma_1^2-\sigma_2^2+\sigma_1.\sigma_2] \leq (S_{yt}/F)^2$

The theory works effectively for ductile material and gives economical fail-safe designs but not under hydrostatic stress state. A comparison of maximum distortion energy and maximum shear stress theories is shown in Fig. 3.12.

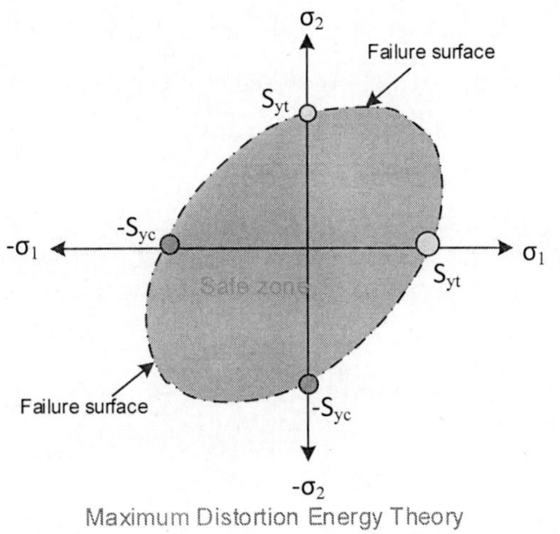

Figure 3.11 Schematic diagram shows failure surface and safe zone considering two normal stress as per Maximum Distortion Energy Theory.

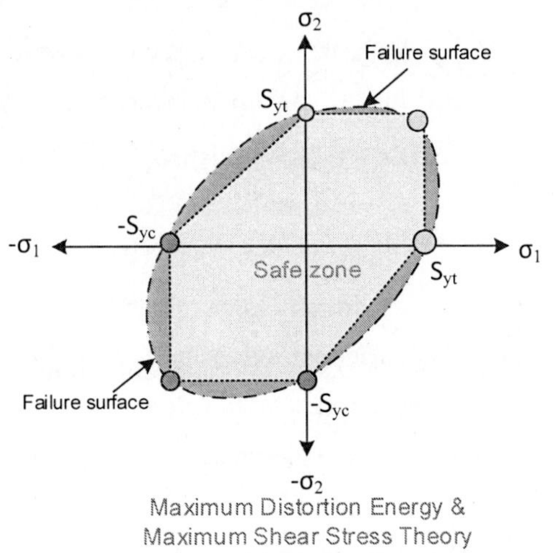

Figure 3.12 Schematic diagram shows comparative failure surface and safe zone corresponding to Maximum Distortion Energy Theory and Maximum Shear Stress Theory.

3.5.4 Coulombs-Mohr Failure Theory

In addition, the maximum principal stress theory, Coulombs-Mohr failure theory is the one suits for brittle materials. The materials showing limited ductility and almost negligible deformation (without any sign prior to fracture) can be categorised as brittle materials. Those brittle materials exhibiting almost the same strength under tension and compression care called even materials (Fig. 3.12) while those showing a significant difference in strength under tension and compression care called even materials are terms as odd materials. Usually, brittle materials show higher strength in compression than tension (Fig. 3.13).

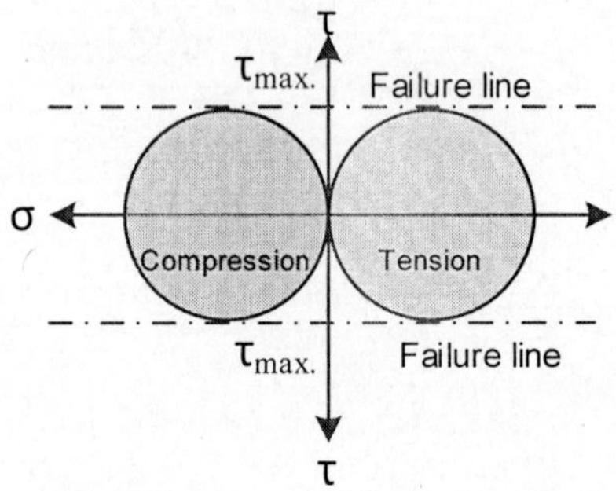

Figure 3.13 Schematic diagram shows Mohr circle for normal and shear stress acting on a stress element of even brittle material.

Brittle material results in a flat fracture surface normal to the tensile loading, while under torsional loading (causing pure shear), the slant fracture surface (45°) is obtained because the maximum normal stress (under torsion) is induced at 45° with respect to the shear loading. The slat fracture surface produced under torsional loading can be similar to ductile failure under tensile loading, however, with distinction with regard to the deformation.

Mohr-Coulomb (MC) theory describes the behaviour of hard and brittle material under combined shear stress and normal stress. It determines the failure load as well as the angle of fracture. Mohr's circle here determines principal stresses leading to such a combination of shear and normal stress, and the angle of the plane (μ).

For even materials, failure stress is independent of normal stresses. In the case of odd materials, the compressive strength of the material is higher than tensile strength. A common line tangential to both these circles forms the failure surface.

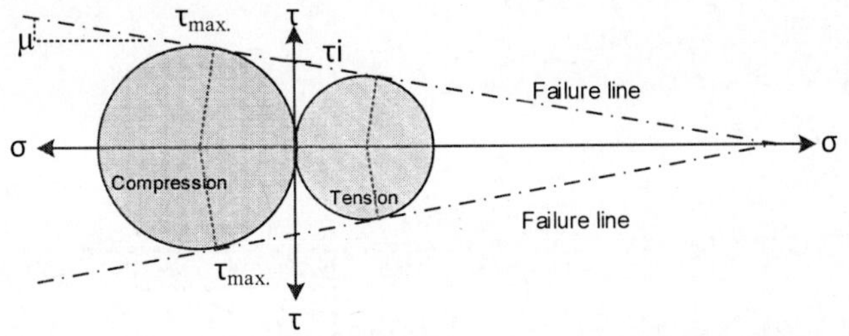

Figure 3.14 Schematic diagram shows Mohr circle for normal and shear stress acting on a stress element of odd brittle material.

As per MC theory, a failure plane is activated by shear stress, wherein the shear stress has a linear relationship with the normal stress component acting on the failure plane. The shear stress and normal stress relation is given as

Shear stress $(\tau) = \tau i + \mu\sigma$

Where σ is the normal stress and τ_i is a constant corresponding to the intercept of failure envelop with τ axis, and μ is the friction angle.

Mohr-Coulomb criterion is the most commonly used shear failure criterion proposed for brittle materials. In this criterion, only the maximum (σ_1) and minimum (σ_3) principal stresses are considered, and it assumes that the intermediate stress (σ_2) is either absent or has no effect (Fig. 3.15 a, b). This criterion ignores the weakening effect of the intermediate stress σ_2 and is considered too conservative in estimating response under tri-axial stress conditions.

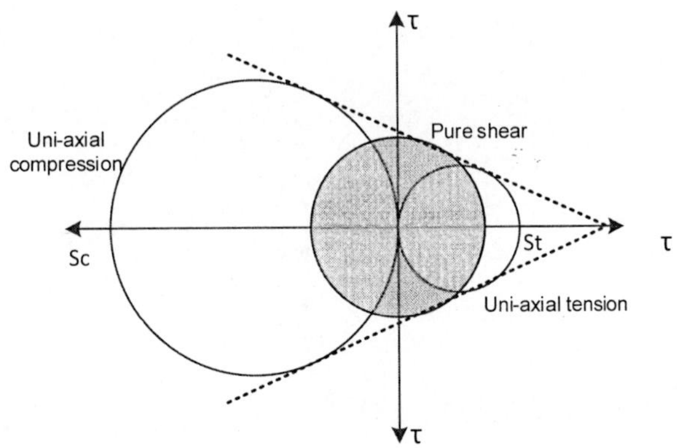

a)

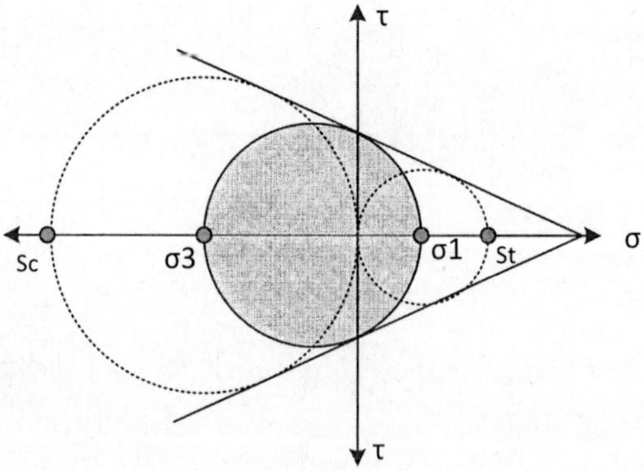

b)

Figure 3.15 Schematic diagram shows Mohr circle with maximum (σ_1), and minimum principal stress (σ_3) acting on a stress element of odd brittle material and their failure surfaces a) schematic representation of uniaxial compression and tensile loading and b) compressive and tensile principal stresses.

According to MC failure theory, the failure surface equation is $\frac{\sigma 1}{St} - \frac{\sigma 3}{Sc} = 1$

Corresponding failure surface generated is as under:

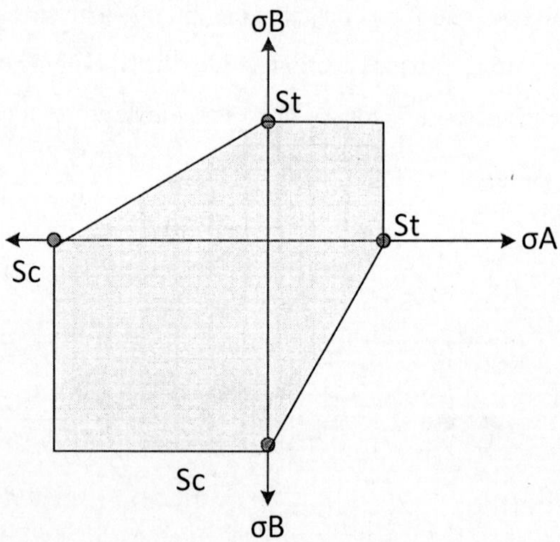

Figure 3.16 Schematic diagram shows failure surface and safe zone as per Mohr-Columbus Theory.

Considering the plane stress condition, let's examine the above mathematical representation for different loading conditions. Say σ_A and σ_B are principal stresses. A combination of principal stresses σ_A and σ_B falling within the green zone of the schematic diagram showing the failure surface will remain safe as per Mohr-Coulombs theory (Fig.

3.16). The σ_A has maximum (σ_1) and minimum (σ_3) principal stress. Under such conditions, there is other influential stress like σ_2.

Examine the situation wherein $\sigma_A \geq \sigma_B \geq 0$, (The first quadrant case)

Plane stress condition, maximum (σ_1) and minimum (σ_3) principal stress.

Means $\sigma_1 > \sigma_B$ and $\sigma_3 = 0$

Therefore, the failure surface equation $\frac{\sigma 1}{St} - \frac{\sigma 3}{Sc} = 1$ is reduced to $\frac{\sigma A}{St} - \frac{0}{Sc} = 1$ or $\frac{\sigma A}{St} = 1$

For the failure to occur $\sigma_A \geq St$ and to avoid failure $\sigma_A < St$

Examine the situation wherein $\sigma_A \geq 0 \geq \sigma_B$, (The fourth quadrant case)

Plane stress condition, maximum (σ_1) and minimum (σ_3) principal stress.

Means $\sigma_1 = \sigma_A$ and $\sigma_3 = \sigma_B$

Therefore, the failure surface equation $\frac{\sigma 1}{St} - \frac{\sigma 3}{Sc} = 1$ is reduced to $\frac{\sigma A}{St} - \frac{\sigma B}{Sc} > 1$

For the failure to occur $\frac{\sigma A}{St} - \frac{\sigma B}{Sc} > 1$ and to avoid failure $\frac{\sigma A}{St} - \frac{\sigma B}{Sc} < 1$

Examine the situation wherein $0 \geq \sigma_A \geq \sigma_B$, (The third quadrant case)

Plane stress condition, σ_A has maximum (σ_1) and minimum (σ_3) principal stress.

Means $\sigma_1 = \sigma_A = 0$ and $\sigma_3 = \sigma_B$

Therefore, failure surface equation $\frac{\sigma 1}{St} - \frac{\sigma 3}{Sc} = 1$ is reduced to $\frac{0}{St} - \frac{\sigma B}{Sc} = 1$ or $-\frac{\sigma B}{Sc} = 1$

For the failure to occur $\sigma_B \geq - Sc$ and to avoid failure $\sigma_B < -Sc$

3.6 Fracture mechanics

Fracture mechanics is a relatively new approach to designing mechanical components. This approach is considered more realistic and practical, especially with regard to the presence of defects and discontinuities in the real materials used to design the mechanical components instead of making assumptions like the material is sound and free from discontinuities. Fracture mechanics establishes the relationship between the loading condition, material properties, and discontinuities present in the material. Fracture toughness is one of the material properties used in the fracture mechanics approach to design the mechanical component. The type and magnitude of stress (σ) induced as per the service load condition and allowable crack-like defect size (c) were included while designing the mechanical component using the fracture mechanics approach (Fig 3.17). Fracture toughness shows the resistance to crack propagation. The fracture toughness of a material is influenced by temperature, environment (e.g., air, fresh water, salt water, chemical, corrosive, etc.), loading rate, material thickness, material processing, and crack orientation to the grain direction. Three parameters are commonly used to measure fracture toughness: stress intensity factor, crack opening displacement, and J integral. As per the ductility, tensile strength, and section thickness, a suitable parameter of the fracture toughness is selected for the design. The stress intensity factor (K) is an appropriate measure of the fracture toughness of the materials (under plane strain conditions) of very limited ductility, high tensile strength, and thick/heavy cross-sections. While other two parameters, crack-opening displacement and J integral, are used to measure the fracture toughness materials (under plane stress conditions) of high ductility, low tensile strength, and thin cross-sections.

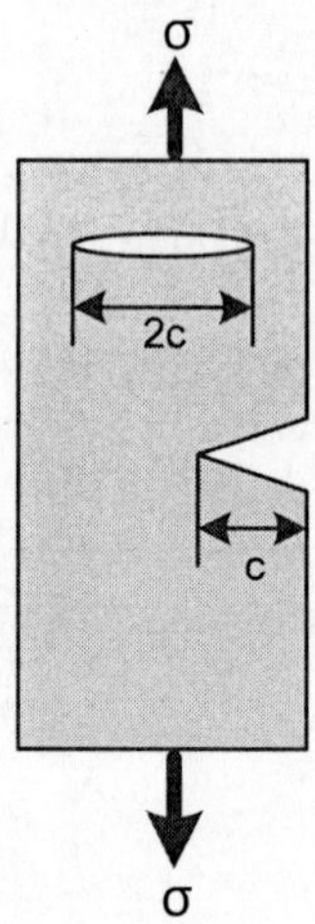

Figure 3.17 Schematic diagram showing open and closed crack in brittle material subjected to tensile stress

The fracture toughness $K = \alpha.\sigma\ (\pi.c)^{1/2}$

Where K is the materials property say critical stress intensity factor, and α is a parameter to consider part geometry and taken as 1.0 for the infinite width of the plate, σ is applied stress, and c is the crack length. In open cracks, c is the length of crack while it is half crack length in closed cracks.

3.6.1 Mode of fracture

In the presence of crack-like defects, the crack propagation depends on the type of stress the crack propagation can occur in three different ways i.e. mode I fracture, mode II fracture, and Mode III fracture (Fig. 3.18). Crack propagation mostly occurs by mode I in the presence of tensile stress wherein the crack faces move apart in a direction perpendicular to the loading direction while the crack mouth is widened. Crack propagation by mode II and mode III occurs due to the shear stress in which crack faces move almost parallel to each other.

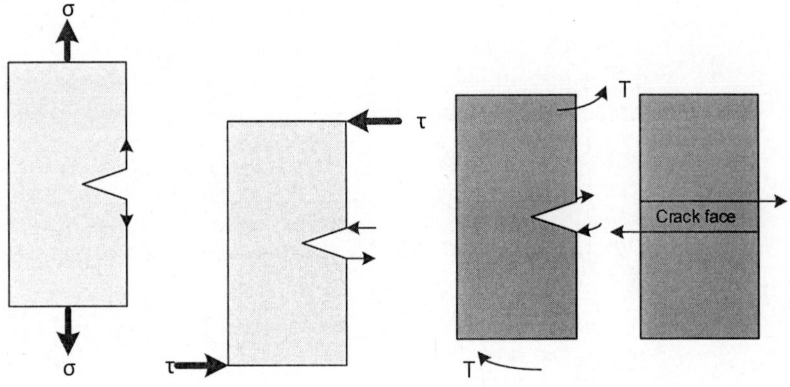

Figure 3.18 Schematic diagram showing the three most common modes of fracture a) Crack opening mode b) In plane shear mode and c) out of plane shear mode

3.6.2 Stress intensity factor and fracture toughness

A component with a crack like defect is subjected to applied tensile stress (σ), stress field in vicinity of the crack tip is developed. Depending on the location of point interest near the crack tip, tensile stress and shear stress fields develop (Fig.3.19). The stress field around (for mode I loading under linear elastic stress state) the crack at a location expressed in polar coordinates can be expressed using the following three equations.

$$\sigma x = \sigma\sqrt{\frac{c}{2r}}\cos\frac{\theta}{2}\left[1 - sin\frac{\theta}{2}.sin\frac{3\theta}{2}\right] + \cdots$$

$$\sigma x = \sigma\sqrt{\frac{c}{2r}}\cos\frac{\theta}{2}\left[1 + sin\frac{\theta}{2}.sin\frac{3\theta}{2}\right] + \cdots$$

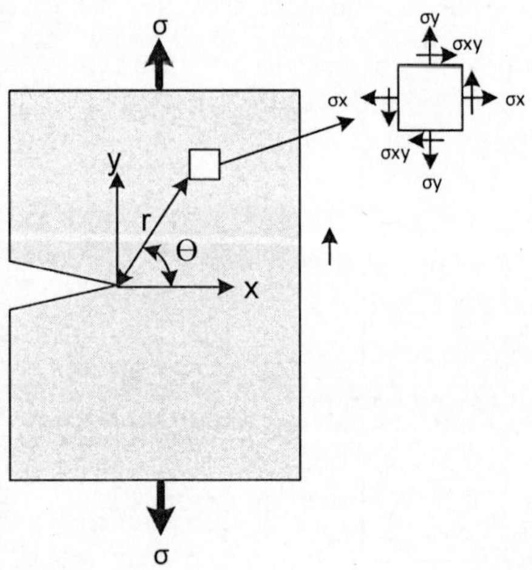

Figure 3.19 Schematic diagram showing an open crack in high strength and low ductility material subjected to tensile stress with stress element near the crack tip

The stress intensity factor (K) shows the stress field around the crack tip and for crack opening mode, I is expressed as $K_I = \sigma\sqrt{\pi c}$ where σ applied stress in MPa and c crack length in mm. A critical combination of stress and crack length results in crack growth causing fracture. K_I for a material decreases with the increase of thickness, and then it becomes constant (Fig. 3.20).

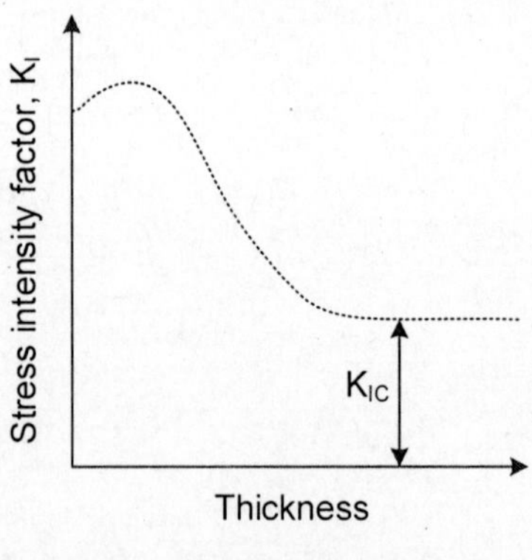

Figure 3.20 Schematic diagram showing variation in stress intensity factor in crack opening mode (K_I) as a function of section thickness

This constant value of K_I of a material, which is independent of section thickness, is called critical stress intensity factor (K_{Ic}) is a material property. The K_{Ic} a material property called fracture toughness used for design of mechanical components.

Fracture stress for materials under plain strain conditions using fracture toughness can be expressed as under $\sigma = \frac{K1c}{\sqrt{\pi c}}$

The stress field near the crack tip (σ_x, σ_y) in terms of fracture toughness and crack size can be written as

$$\sigma x = \frac{K}{\sqrt{2\pi r}}\cos\frac{\theta}{2}\left[1 - sin\frac{\theta}{2}.sin\frac{3\theta}{2}\right] + \cdots$$

$$\sigma y = \frac{K}{\sqrt{2\pi r}}\cos\frac{\theta}{2}\left[1 + sin\frac{\theta}{2}.sin\frac{3\theta}{2}\right] + \cdots$$

To determine the stress field (σ_x, σ_y), just the head of the crack tip the value of θ will be zero. The tensile stress σ_y (in loading direction) causing the crack growth is reduced to $\sigma y = \frac{K}{\sqrt{2\pi r}}$

This concept of estimating the tensile stress σ_y (in loading direction) leading crack growth can be re-written as $\sigma y = \frac{K}{\sqrt{2\pi r}}$

This shows that the tensile stress (in the direction of loading) ahead of the crack tip decreases with distance from the crack tip (Fig. 3.21). This stress is expected to cause crack propagation and fracture therefore, it is important from the design point of view. Further, this equation suggests that tensile stress σ_y at the crack tip (where the value of r is zero) will be infinite, and no material should stand under such conditions.

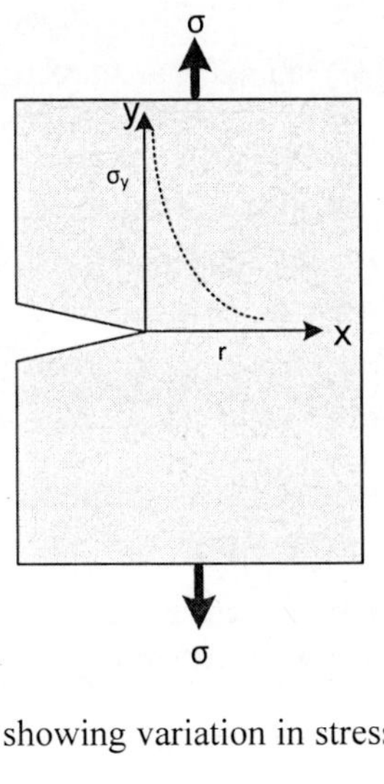

Figure 3.21 Schematic diagram showing variation in stress (σ_y) as function of distance (r) from crack tip

3.6.3 Plastic zone size

The real metals, however, experience plastic deformation near crack and the tip does not remain sharp due to blunting of the crack tip (Fig. 3.22). Plastic zone size near the crack tip for a given stress, crack and fracture toughness can be obtained.

Stress in loading direction $\sigma y = \frac{K}{\sqrt{2\pi r}}$ = yield strength (Sy)

$Sy^2 = \frac{K^2}{2\pi r}$ so $r_p = \frac{K^2}{2\pi\, Sy^2}$

Plastic zone size (r_p)= $\frac{1}{2\pi}(\frac{K}{Sy})^2$

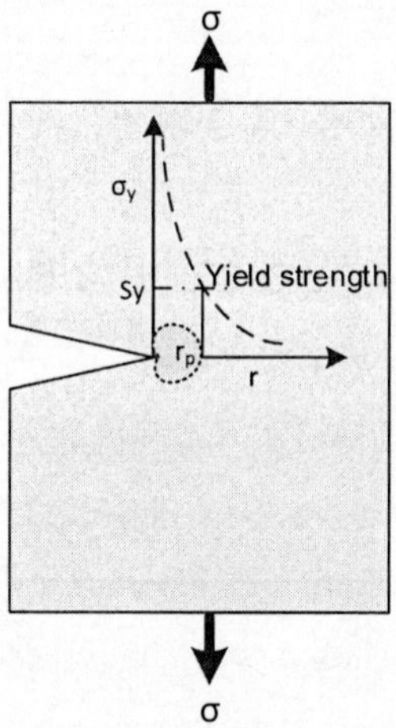

Figure 3.22 Schematic diagram showing plastic zone formation near the crack tip

For plain strain conditions, the requirement of a section thickness of material (t) for given tensile strength should be satisfied as under:

Minimum thickness (t_m) $\geq 2.5\ (K_{IC}/Sy)^2$

If above requirement is not satisfied, then fracture toughness K_c of the material for a given thickness can be approximated using the following equation $K_C = K_{IC}\,[1 + B\, e^{-(\frac{At}{tm})^2}]$

Where A and B are material constant, and t_m is the minimum material thickness for plain strain conditions.

3.7 Griffith Theory on Brittle Fracture

The Classical Griffith Criterion of fracture is applicable for brittle materials. It is used that all realistic materials are not perfectly sound and have crack-like discontinuities. These cracks grow catastrophically under the influence of tensile loading, and cracks grow without any plastic deformation near the crack tip. The growth of the crack is accompanied by the breaking of the atomic bonds ahead of the crack tip leading to the creation of two new crack surfaces. The energy is released due to breaking the (elastic) bonds of the atoms while energy is consumed to create new surfaces. The former energy is known as the "energy released and is considered negative (as it is lost), while the latter one is termed surface energy. The surface energy is taken as a positive and is used to create new surfaces. Released energy must be enough to provide the required surface energy to create the new surface because of the crack growth.

According to this theory, the necessary condition for brittle fracture is that amount of energy released due to the breaking of atomic bonds of material ahead of the crack tip is at least equal to the amount of energy needed for the creation of the new surfaces generated by cracks (Fig. 3.23). Conversely, the energy released should be more or equal to surface energy. Considering γ_s surface energy for creating unit surface area, E modulus of elasticity, c crack length and σ is applied tensile stress, the stress required to cause brittle fracture can be obtained as under.

The energy released (Ur) = $-\frac{\pi\sigma^2 c^2}{E}$

Surface energy (Us) = $4c\gamma_s$

Total energy (U)=Energy released (Ur) + Surface energy (Us) = $-\frac{\pi\sigma^2 c^2}{E} + 4c\gamma_s$

As per this theory, the minimum requirement for crack growth is that the total energy due to crack growth should be zero.

$$\frac{dU}{dc} = \frac{d}{dc}\left[-\frac{\pi\sigma^2 c^2}{E} + 4c\gamma_s\right]$$

$$= -\frac{2\pi\sigma^2 c}{E} + 4\gamma_s = 0$$

$$\frac{2\pi\sigma^2 c}{E} = 4\gamma_s$$

$$\sigma^2 = \frac{2\pi E\gamma s}{c}$$

Stress to cause brittle fracture $\sigma = \sqrt{\frac{2\pi E\gamma s}{c}}$

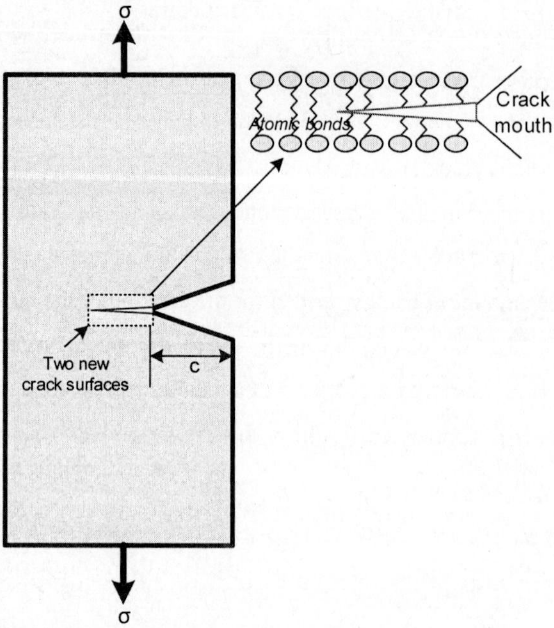

Figure 3.23 Schematic diagram showing crack growth of an open crack under tensile stress

Modified Griffith Theory

However, this theory assumes that plastic deformation near the crack tip is absent and the crack tip remains sharp without any blunting. In real materials, there is always some elastic/plastic information near the crack tip that may be at the micro level. Some energy is consumed to cause this micro-scale plastic deformation in brittle materials (Fig. 3.24). This increases the requirement of energy for the growth of cracks and the creation of surfaces. The classical Griffith theory was modified to accommodate the possibility of microscope deformation at the crack tip by including the additional energy required for plastic deformation (γ_p). Then, the stress required to cause fracture increases significantly and may be written as:

$$\sigma = \sqrt{\frac{2\pi E(\gamma p+\gamma s)}{c}}$$

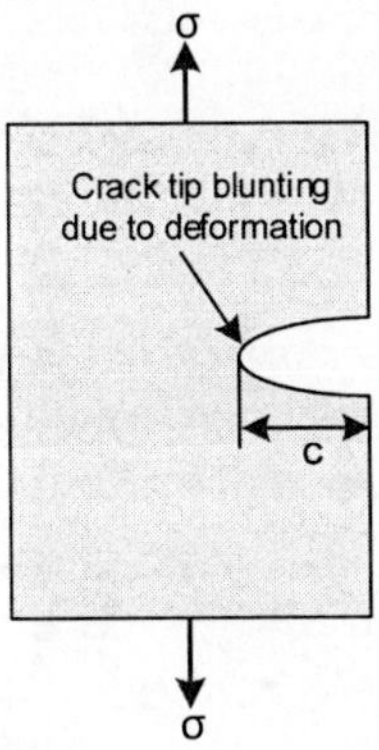

Figure 3.24 Schematic diagram showing blunting of crack tip (increased crack tip radius) due to micro-scale plastic deformation

It has been observed that the energy for plastic deformation (γ_p) is significantly greater (100 to 1000 fold) than the surface energy (γ_s). Then above equation can be further simplified as under.

$$\sigma = \sqrt{\frac{2\pi E\gamma p}{c}}$$

Any approach to reduce the surface energy required for the creation of the new surfaces resulting from crack growth helps reducing the stress required for brittle fracture for cutting application.

3.8 Fatigue failure

The mechanical components during the service experience different types of load. Service load can be static or dynamic. The magnitude and the direction of the static load largely remain constant. The magnitude and the direction of the load change during the service under the dynamic loading. Depending upon the fluctuation in load and its nature, different dynamic loading patterns are observed (Fig. 3.25). The extent of fluctuation in load and their type determines the life and performance of a mechanical component significantly. Fluctuating loads can also be terms as fatigue loads. However, the fluctuation in both load magnitude and direction must be high enough to cause fatigue. In general, the mechanical components under fluctuating loads perform poorly than static loads. The mechanical performance and load carrying capacity are drastically compromised under the fatigue loading in the presence of the stress raisers. The time for which a component service under the fatigue loading is called fatigue life. Fatigue life is generally characterised by the number of load cycles it takes before fatigue fracture. The fatigue life of components is severely compromised in the presence of stress raisers, especially in hard and brittle materials.

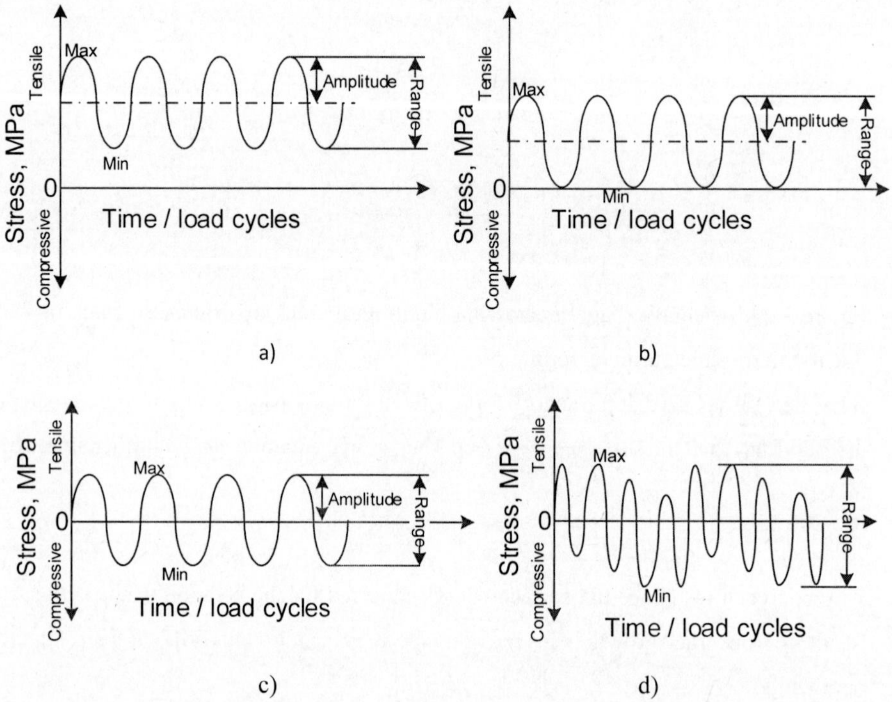

Figure 3.25 Schematic diagram showing different types of fatigue loading a) tensile-tension with regular pattern, b) tensile-zero with regular pattern, c) tensile-compression with a regular pattern, and d) tensile-compression-zero with an irregular pattern called fluctuating load

Common terms related to fatigue loading include maximum stress, minimum stress, stress range, mean stress, and stress amplitude. These are obtained from the load vs time curve using the following relationship.

$$Mean\ stress\colon \frac{\sigma \max + \sigma min}{2}$$

$$Stress\ range\colon \sigma \max - \sigma min$$

$$Stress\ amplitude\colon \frac{stress\ range}{2}$$

$$Stress\ ratio\colon \frac{\sigma min}{\sigma max}$$

The sign convention for tensile stress (+) and compressive stress (-) and accordingly these should be used for calculating the mean stress, stress range, stress amplitude and stress ratio. A low value of stress ratio causes more fatigue than a high value as load tends to behave more like static load with very little fluctuation.

The fatigue fracture of a smooth component occurs in three steps namely crack nucleation by slip, stable crack growth, and sudden fracture as shown in Figure. These three stages determine the number of load cycles a component will take for fatigue fracture. Each stage of the fatigue fracture needs a different number of load cycles. The crack nucleation stage and stable crack growth stage account for 80 to 90% of the useful fatigue life. The remaining 10-20 % of the fatigue life is consumed by the sudden fracture stage (Fig. 3.26). The component is usually taken off from the service as soon as it reaches the third stage of the fatigue fracture if detected timely otherwise, catastrophic fracture is observed. Attempts are always made to design and manufacture the components such that the crack nucleation and the stable crack growth stages are delayed. The crack nucleation stage is delayed by making the surfaces smooth and free of stress raisers of reasonably high hardness and strength by resisting slip and micro-scale surface deformation leading to crack nucleation. A good combination of ductility, strength, and fracture toughness of material help to delay the stable crack growth stage by blunting the growing crap tip, reducing the crack growth rate, and increasing the critical crack size needed for sudden fracture of the third stage.

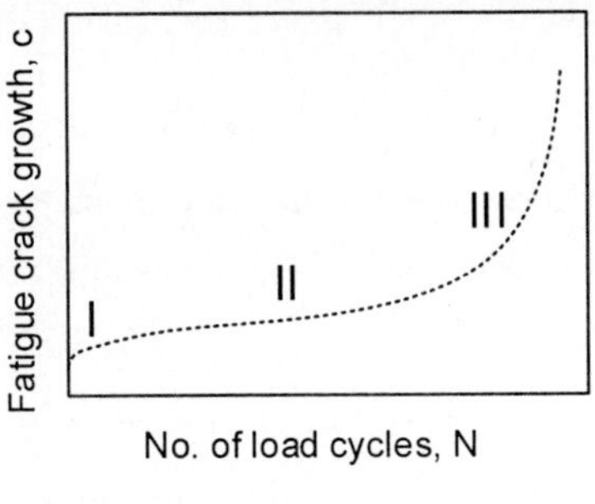

Figure 3.26 Schematic diagram showing the fatigue crack growth as a function of number of fatigue load cycles and stages of fatigue fracture

3.8.1 Crack nucleation

Continuous fluctuation, e.g. in the tensile-compressive load, causes the maximum shear stress at 45°. Shear stress results in the slip at the plane of maximum shear stress. Under the tensile load, slip occurs in one direction, while under the compressive slip takes place in the opposite direction. Continuous fluctuation in magnitude and the direction of the loads develop many cracks like micro-scale surface irregularities (Fig. 3.27). These micro-scale irregularities eventually grow large size enough to act as cracks. The number of fatigue load cycles required to complete the crack nucleation stage depends on all the factors (yield strength, ductility and hardness, stress raisers, and surface roughness) affecting the slip. An increase in hardness, yield strength of the material, and smoothness of the surface increase the number of cycles required to complete the crack nucleation stage (10-15% of fatigue life), which in turn helps to enhance the fatigue performance and fatigue life.

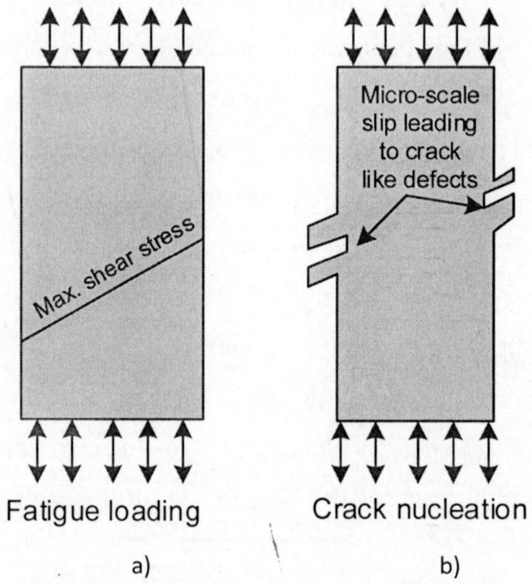

Figure 3.27 Schematic diagram showing fatigue crack nucleation on smooth surface a) the maximum shear stress plane prone to slip and b) micro-cracks (in form of protrusion and cavity) developed because of slip due to fatigue loading

3.8.2 Stable Crack Growth

During the stable crack growth stage II, the crack propagates steadily at a very low rate. The rate of crack growth, however, increases gradually (Fig. 3.28). The rate of crack growth is determined by the service load conditions and the material properties. The high-stress range, maximum stress, and stress amplitude of the fluctuating load cause a higher crack growth rate while fracture tough, and ductile metals blunt the growing crack's tip and increase the stress range/amplitude required to cause the crack growth. Therefore the material structure, mechanical properties affect the fatigue life significantly.

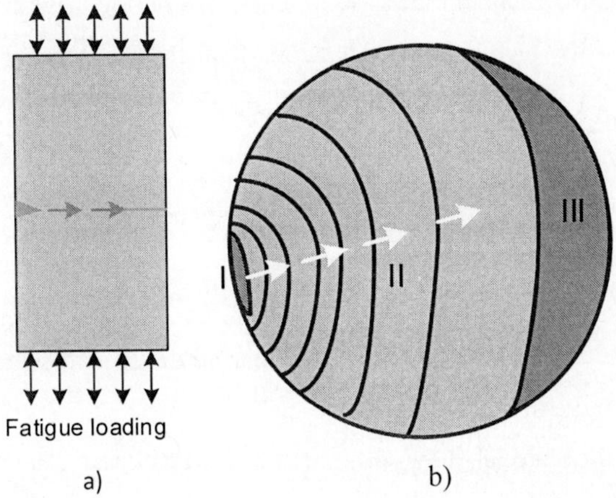

Figure 3.28 Schematic diagram showing stages of fatigue crack and stage II corresponding to stable crack growth a) fatigue crack across section, and b) typical features observed on the fatigue fracture surface in different stages of fatigue

The fatigue crack growth rate is $dc/dN: P(\Delta K)^m$

Where P is crack growth rate when stress intensity factor range ΔK: $[\sigma max - \sigma \min)\sqrt{\pi c}]$ is equal to 1. σ_{max} and σ_{min} are maximum and minimum stress, and c is the crack length. The parameter m is related to the material and indicates the slope of the curve in fatigue stage II (Fig. 3.29).

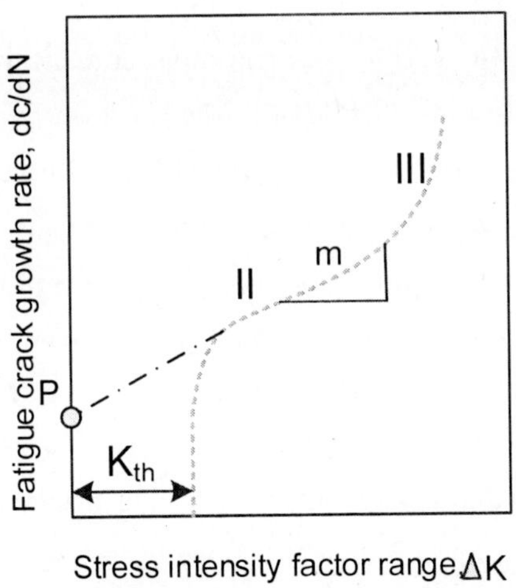

Figure 3.29 Schematic diagram showing stable fatigue crack growth rate as a function of stress intensity factor range

During fatigue crack growth stage, the crack size increases gradually, reducing the load resisting cross-sectional area and increasing the stress range/amplitude for the given load fluctuations. Conversely, an increase in crack size during stage II continuously increases the stress intensity factor range in a stable manner. The threshold stress intensity factor range (ΔK_{th}) is a minimum stress intensity factor range needed to cause the propagation of the existing crack. Suppose the stress intensity factor range (ΔK) is lesser than the threshold stress intensity factor range (ΔKth); then the existing crack becomes a non-propagating type and so fatigue fracture does not occur.

3.8.3 Sudden fracture

As soon as the crack attains a critical size during the second stage of fatigue, the crack growth rate increases exponentially, and sudden fracture is observed. The fracture toughness of the material shows a higher critical crack size. Therefore, the third stage of fatigue fracture is minimised in the case of fracture-tough materials. On the other hand, low fracture toughness materials show a large fracture surface area corresponding to the third stage, catastrophic fracture.

The crack nucleation stage is typically characterized by the presence of distinct colour, texture, and multiple cracks. The stable crack growth stage usually shows a concentric circle similar to beach marks with the centre at the crack nucleation site. The 3rd stage fatigue fracture is typically an overload fracture that may show ductile or brittle surface features.

3.9 High cycle fatigue

The fatigue fracture of material can occur in fewer to a very high number of load cycles depending upon the loading conditions. The fatigue fracture of the material under the elastic stress/ strain conditions takes a very large number of load cycles (1 or 2 million); therefore, it is termed high cycle fatigue. On the other hand, fatigue fracture of the material takes place in very few load cycles (<10,000) under the stress/strain loading conditions beyond the elastic limit accordingly, it is termed low cycle fatigue. The mechanical components like a shaft, connecting rod, chassis, etc. experience are subjected to high cycle fatigue. Components in thermal power plants subjected to varying thermal cycles (heating and cooling from temperatures like 500 °C or more to room temperature) frequently experience low cycle fatigue. Therefore, different approaches to designing the mechanical component are used for low and high-cycle fatigue. The high cycle fatigue test of the material is conducted for load cycles like 10,00,000 or 20,00,000 as per the requirement of the application using suitable fatigue stress conditions.

The high cycle fatigue behaviour of a material is usually expressed using the S-N curve. S represents the stress component (maximum stress/stress range /stress amplitude), and N shows the number of fatigue load cycles. In general, a reduction in the fatigue stress component (maximum stress/stress range /stress amplitude) up to a limit increases the number of load cycles required to cause the fatigue fracture. Thereafter, fatigue life (at further lower stress) becomes independent of the fatigue loading conditions. This behavior suggests that to cause the fatigue fracture of a component, two conditions must be satisfied a) sufficient load/stress fluctuations to cause a reasonable stress range (else component's response will be like under static loading) and b) the maximum stress must be high enough to facilitate the crack nucleation and its growth during the fatigue.

3.9.1 S-N Curve and fatigue strength

Two types of S-N curves are commonly observed as shown in Fig. 3.30. S-N curve wherein metals like carbon steel showed an increase in fatigue with the reduction in the maximum/ stress range/amplitude up to a limit thereafter, the fatigue life becomes independent of the fatigue loading conditions. This maximum stress/range/ amplitude below which the fatigue fracture is not observed is called the endurance limit. The endurance limit suggests that lower fatigue stress conditions will offer infinite fatigue life. The endurance limit is a material property used to design the mechanical component against high cycle fatigue. This may also be termed as fatigue strength of the material. The endurance limit for steel and copper alloys is found to be in a range of 0.35 to 0.5 time tensile stress. For example, alloy steel's fatigue strength (stress range, amplitude, maximum stress) is 250 MPa for 1000000, 2000000 load cycles or more.

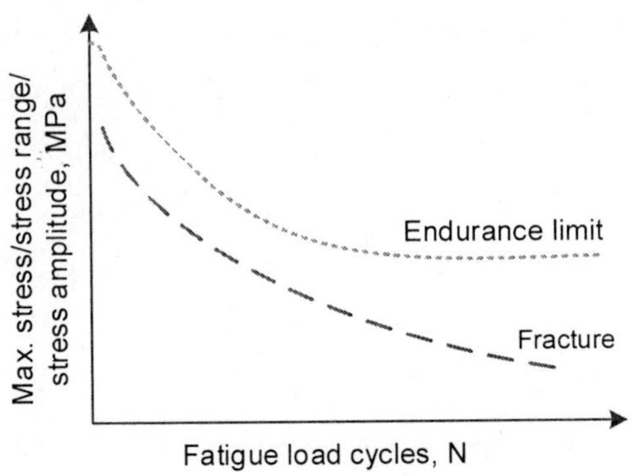

Figure 3.30 Schematic diagram stress (S) vs. number of fatigue load cycle (N) curves for two different types of metals

A different approach is used to determine the fatigue strength of materials (for design purposes) showing continuous increasing fatigue life with the decrease in the stress range / maximum stress and stress amplitude.

In this case, maximum stress/range/ amplitude is obtained from the S-N curve for a specific number of load cycles for which a component is to be designed (Fig. 3.31). The fatigue strength of such materials, therefore, is expressed for specific fatigue life. For example, the fatigue strength (stress range, amplitude, maximum stress) of metals is 100 MPa for 500000 load cycles.

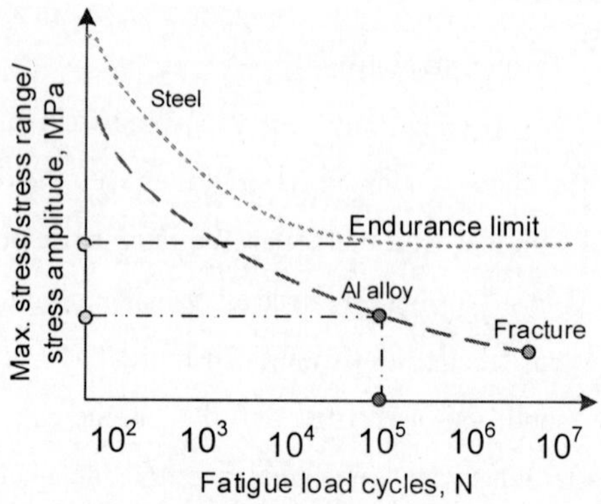

Figure 3.31 Schematic diagram showing the approach to estimate the fatigue strength of two different types of metals

3.10 Stress-life approach

Mechanical components can be subject to a variety of fluctuating loads. If the pattern of load fluctuation remains constant, then the above-described approaches can be easily used for fatigue-resistant design by establishing endurance limit or fatigue strength. However, suppose the magnitude of the fluctuating loads changes over a period as shown schematically in the figure 3.32. In that case, the cumulative fatigue damage approach is used to establish the fatigue life. The sequence of load fluctuations does not affect the fatigue life.

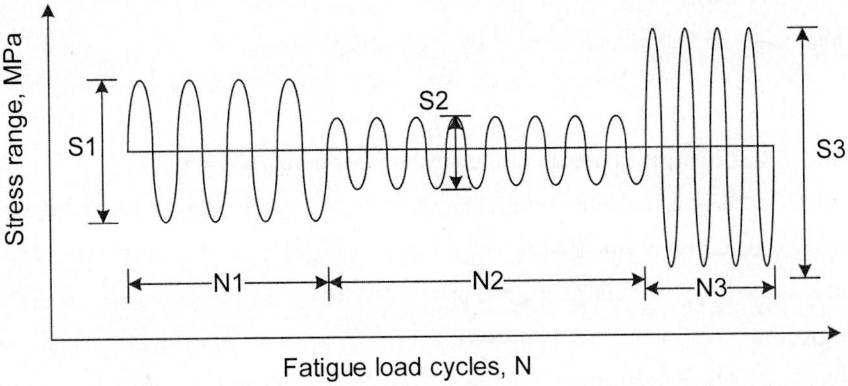

Figure 3.32 Schematic diagram fluctuating stress range (S1, S2, S3) for different number of fatigue load cycles (N1, N2, N3)

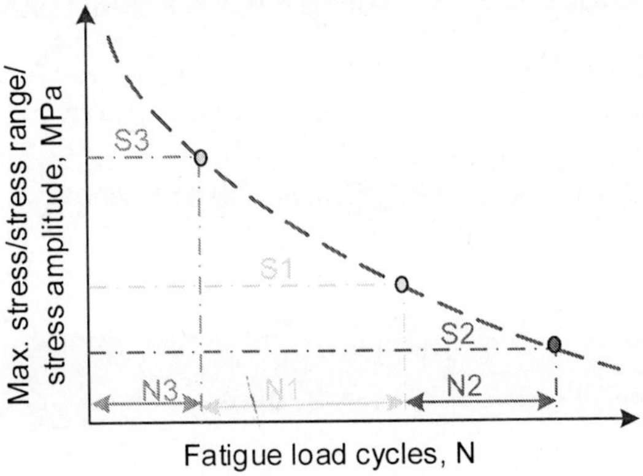

Figure 3.33 Schematic diagram cumulative damage approach when a component is subjected to different stress ranges / amplitudes (S1, S2, S3) for varying number fatigue load cycle (N1, N2, N3) curves

The fatigue life of a component in terms of number of load cycles is N for a given stress range/maximum stress. If the same component is subjected different maximum stress/stress ranges (S1, S2 and S3) for varying number of fatigue load cycles (N1 N2, N3). Each set of stress and load cycle will be causing damage in the form of crack growth. To cause the fatigue fracture, the damage (crack growth) is taken cumulative (Fig. 3.33). Assuming that sequence loading (S1, S2 and S3) for varying load cycles do not have any impact on fatigue life, the exposure to different load cycles (N1, N2, N3) at varying stress conditions and total fatigue life (N) can be related using the following relation.

$$\frac{N1}{N} + \frac{N2}{N} + \frac{N3}{N} + \cdots = 1$$

3.11 Effect of mean stress on fatigue using Goodman diagram

The most common method of conducting the fatigue test of material is rotary fatigue testing using a load cycle of equal tensile and compressive stress with stress ratio (R) -1 and means stress (σ_m) 0. The fatigue strength is obtained in terms of stress amplitude (σ_a) for a given fatigue life (N, number of load cycles). To develop S-N curve similar fatigue tests are conducted at reducing stress amplitude / stress ratio / maximum stress and corresponding number of load cycle for fatigue fracture are obtained.

Using the given data of fatigue strength (σ_a), obtained from fatigue test using mean stress (σ_m) 0, the fatigue strength for the design purpose under another set of service conditions (σ_m) and tensile properties of metals can be estimated using three following three established equations (Fig. 3.34).

Fatigue strength (σ_a) for a new load condition as per Soderbeg: $(\sigma_a)_{\sigma m:0}\ [1 - \frac{\sigma m}{\sigma y}]$

Fatigue strength (σ_a) for a new load condition as per Goodman: $(\sigma_a)_{\sigma m:0}\ [1 - \frac{\sigma m}{\sigma u}]$

Fatigue strength (σ_a) for a new load condition as per Gerber: $(\sigma_a)_{\sigma m:0}\ [1 - (\frac{\sigma m}{\sigma y})^2]$

Where σ_y and σ_u are the yield and ultimate strength stress of metal. σ_m is the mean stress. $(\sigma_a)_{\sigma m:0}$ is the fatigue strength in terms of stress amplitude from the rotary fatigue test, which means stress σ_m= 0.

Soderberg give very conservative estimate of fatigue strength while Gerber equation given reasonable good results for ductile metals.

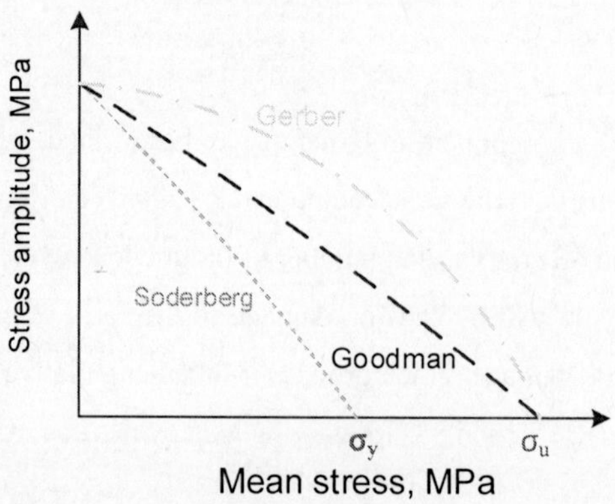

Figure 3.34 Schematic diagram showing variation in fatigue strength in terms of stress amplitude as a function of mean stress according to Soderberg, Goodman and Gerber equations

3.12 Fracture with fatigue

The fatigue fracture usually happens abruptly without any prior indication. The fatigue fracture surface is generally flat and does not show any prior indication like plastic deformation in the vicinity of the fracture surface (Fig.3.35). Therefore, it can be disastrous in terms of loss of time and property. Therefore, it is important that the critical mechanical component is carefully examined during the service for the presence of any fatigue crack and its growth using various non-destructive testing / examination (NDT/ NDE) approaches like vibration measurement, ultrasonic testing etc. Cracks present in a mechanical component may be non-propagating or propagating type. Only the propagating type of fracture cracks needs attention and regulation examination for their growth. As soon as a propagating fatigue crack attains size close to the critical crack size either a component is taken off the service or necessary repair/refurbishing work is done.

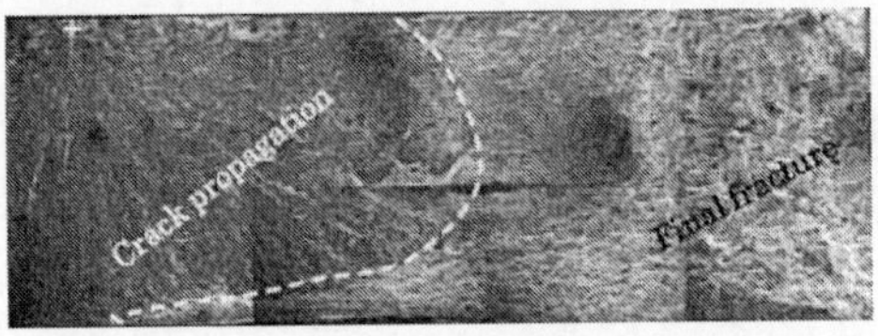

Figure 3.35 Scanning electron micrograph of fracture surface showing three stages of fatigue fracture (Sharma et al. 2014)

3.13 Non-Destructive Testing

The assessment of the soundness of the material and the component during the manufacturing and service is required. Assessment of the material during the manufacturing is done to detect the defects and discontinuities to avoid further processing and manufacturing of defective products and take suitable corrective measures. The evaluation of the components during the service is primarily done to detect the presence of any crack and its growth. Evaluation of the component during service is done using non-destructive techniques. The examination of the material at the manufacturing is done using both destructive and non-destructive techniques. Destructive techniques help to establish the load-carrying capacity of the material under different types of static and dynamic loading conditions. Such characterisation helps to take a suitable decision regarding the suitability of the material for a particular application, the design of the components, and the selection of suitable manufacturing processes to make the product. Material subjected to the destructive test is broken and does not remain useful for any other purpose therefore, it is called destructive testing. While in non-destructive testing, the material is not subjected to any damage, and therefore the functionality and performance of the component, even after the non-destructive test is not compromised.

The defects and discontinuities detected using the non-destructive techniques are broadly grouped into two categories: surface defects and internal defects. Internal defects can be near the surface (sub-surface defects) or deep below the surface, like porosity, inclusions, internal cracks, piping defects, etc. It is considered easier to detect surface defects than sub-surface defects. Surface defects are detected using the techniques like dye penetrant test, magnetic particle test, and eddy current test. Near-surface defects (sub-surface below 1-3 mm) can be detected using magnetic particle test, eddy current test, and radiographic test. The defects and discontinuities at a greater depth below the surface are examined using the ultrasonic test and radiographic test.

3.13.1 Dye Penetrant Test

The dye penetrant test is relatively easier and cost-effective. However, this test is used to detect surface defects only. The surface defects like cracks, open pores, crevices, and pits can be detected in terms of size, shape, and location. The Dye penetrant test is conducted in four steps: surface cleaning, application of the dye over the surface to be examined, cleaning of the dye from the surface, and application of the developer. The stain formed on the developer indicates the size, shape, and location of discontinuities. These four steps are schematically shown in the figure (Fig. 3.36).

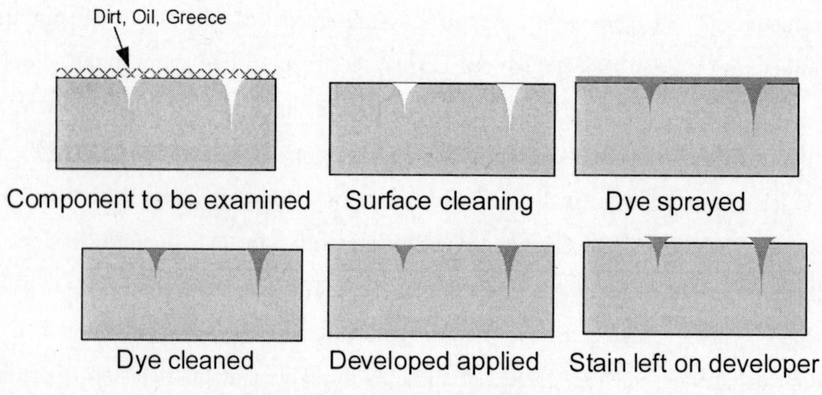

Figure 3.36 Schematic diagram showing non-destructive testing by dye penetrant method using steps like cleaning, dye application, dye cleaning, applying developer and stain observations

3.13.2 Magnetic Particle Test

A magnetic particle test is primarily conducted on ferromagnetic materials to detect the presence of surface and near-surface defects. The magnetic particle testing is conducted in three steps i.e. magnetising the component to be examined, sprinkling the magnetic powder particles over the magnetised surface, and observing the pile-up pattern of powder particles over the surface. The component to be tested is magnetised suitably using the electromagnetic field or permanent magnets. Mostly the components are magnetised using the electromagnetic principle. Once the component is magnetised, the magnetic flux leaks from the location whenever the defects at the surface and near the surface layer are present and create two additional poles (N & S) as shown in Fig. 3.37. Thereafter, a fine powder of the magnetic material is sprinkled over the magnetised surface of the component to be examined. Powder particles of magnetic materials are attracted toward the location of the leaking magnetic flux. Accordingly, the magnetic powder particles piled up. The pattern of piling up of the magnetic particles over the surface indicates the location, size, and shape of the discontinuities. The piling up of the powder of articles is clearer and more intense in case of surface discontinuities than the subsurface defect. The uniform distribution of the magnetic powder particles over the surface indicates the absence of any discontinuity.

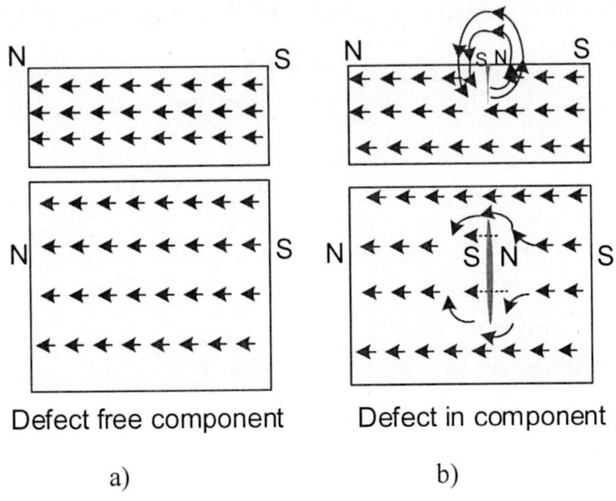

Figure 3.37 Schematic diagram showing non-destructive testing by magnetic particle testing a) flow of magnetic lines of flux in ferromagnetic material free from any defects and b) disturbance in the flow of magnetic lines of flux in the presence of defects

3.13.3 Ultrasonic testing

Ultrasonic testing is one of the most commonly used non-destructive techniques to examine internal defects. This technique helps to establish the size and location of the defects without causing any damage to the surface of the component. However, it needs the expertise to undertake ultrasonic tests and interpret the significance of the results. Ultrasonic waves have the capability to penetrate and travel through different mediums, including air, metal, and non-materials. These waves are, however, reflected back from the interfaces whenever there is a change in medium.

This behavior of the Ultrasonic waves is used to assess the presence of defects internally. These waves are generated using a suitable transducer that converts the electrical current of ultrasonic frequency into ultrasonic vibration. Sound component free from any defect and discontinuity shows two peaks of reflected waves, i.e. one from the top (air-metal internal) and another from the bottom (metal-air interface), as shown in Fig. 3.38. The presence of any defect between the top and bottom surfaces of the component being examined shows an additional peak (in between the two earlier peaks). The surface area interest of the component is scanned using the suitable probe to examine the soundness. The relative position of the intermediate peak indicates the location of the defects between the top and bottom surfaces (Fig. 3.39). The ultrasonic wave reflection is very intense and strong from the defects oriented perpendicular to the path of ultrasonic waves; while defects oriented parallel to the direction of movement of ultrasonic waves may not be detected easily.

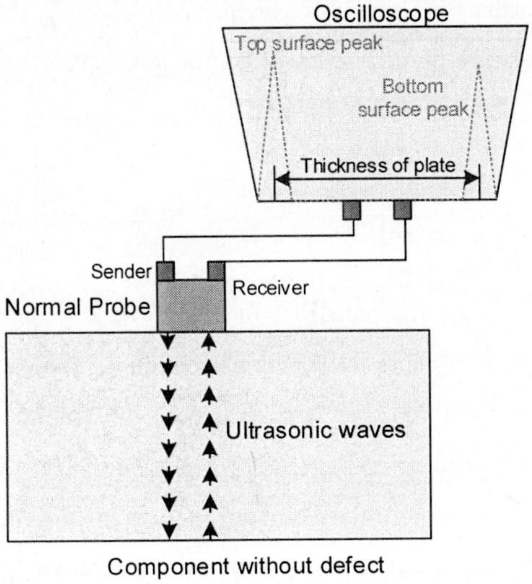

Figure 3.38 Schematic diagram showing non-destructive testing by ultrasonic testing showing two peaks recorded in oscilloscope corresponding to top and bottom interfaces of a sound metal

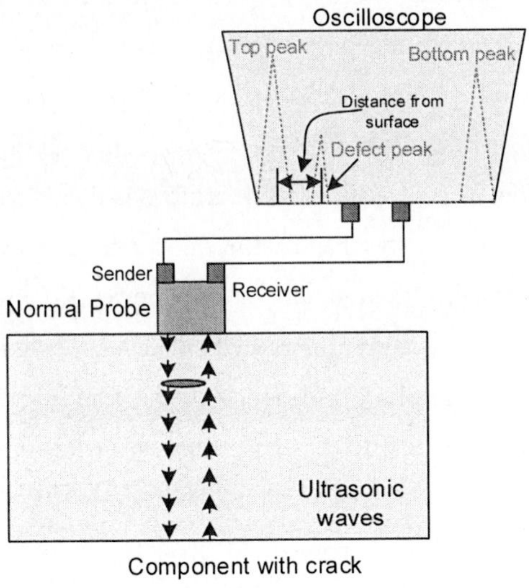

Figure 3.39 Schematic diagram showing non-destructive testing by ultrasonic testing with an additional third peak in between two (top and bottom) peaks in the oscilloscope, which corresponds to the location of the defect

Ultrasonic testing probes available nowadays act as both source (sender) and receiver of the (reflected) ultrasonic vibration. These probes can transmit ultrasonic vibrations at 90 degrees, 60 degrees, and 45 degrees. The 90-degree probe, also called a normal probe, is mainly used to measure the thickness of components (plates, sheets) having flat surfaces on both sides. The oblique probes (60- and 45 degree probes) are used to test the soundness of weld joints.

3.13.4 Radiographic testing

Radiographic testing uses the radiations like X-rays and gamma rays to conduct the non-destructive testing of very thin to very thick section components. For extremely critical applications like nuclear reactors and spacecraft, the soundness of the components is examined using the radiographic testing technique. The X-ray technique for non-destructive testing is used for relatively thin sections, while the gamma rays technique is preferred for thick sections. The X-ray technique is relatively faster than the gamma-ray technique of radiography.

The radiographic technique for non-destructive testing is based on the principle of irritation capability of X-Ray and gamma rays and passes through the metals and affects the X-ray films located other side. However, the extent of absorption of these radiations varies with the material. The metals absorb these radiations to a greater extent than the other low-density materials like inclusion, air pockets, and other defects.

The component to be examined is exposed to the X-ray (gamma ray), as shown in Figure. A metallic component free from defect and discontinuity absorbs the X-rays significantly; therefore, X-ray film placed on another side is affected lightly and uniformly over the entire area (Fig. 3.40). On the other hand, a metallic component with pores, cracks, and air pockets, when exposed to the radiation absorbs fewer X-rays at the location where the defects are present and so X-ray film placed on the other side is affected greatly which is reflected in the form of localized dark spots in the X film (Fig. 3.41).

Radiographic techniques help to develop a permanent record of non-destructive testing, which can later be used for reference and training purposes. To locate the defect in a three-dimensional component, it is required to conduct the radiographic test of the same component at a different orientation/direction.

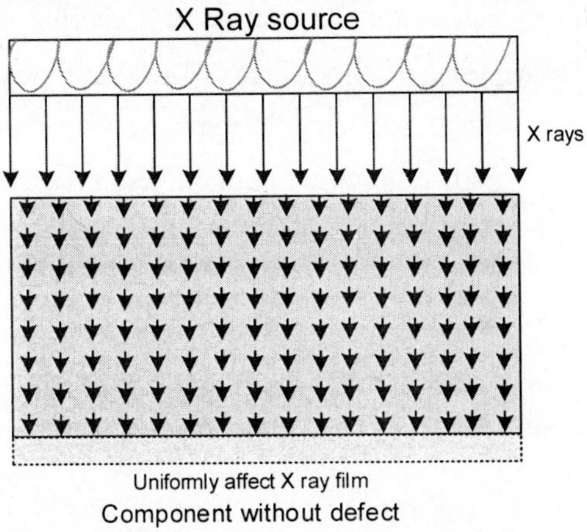

Figure 3.40 Schematic diagram showing X-ray testing where in X rays travel through the metal and affect the X ray film uniformly in case of sound metal

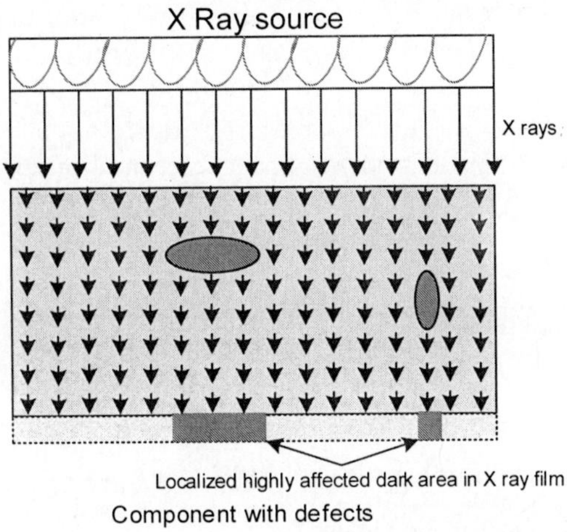

Figure 3.41 Schematic diagram showing X-ray testing where in X rays travel through the different materials on its path and affect the X ray film differently (in different area) in case of metal with defects

UNIT SUMMARY

Mechanism of the ductile and brittle failures has been presented. The need for theories of failure under Static loading has been explained. Common theories of failures under static loading like maximum principal stress theory, maximum shear stress theory, maximum distortion energy theory for ductile materials, and Mohr-Coulombs Theory of brittle materials, have been described. Concept of the fracture mechanics has been introduced. The stress intensity factor as a measure of fracture toughness has been elaborated. Concepts of fatigue life and fatigue strength have been explained using the S-N diagram. The need for non-destructive techniques has been presented. Methodology to examine the surface and subsurface defects using dye penetrant testing, magnetic particle testing, ultrasonic testing, and x-ray testing have been explained using suitable schematic diagrams.

EXERCISE

Multiple Choice Questions

Questions for self-assessment

1. Flat and Square fracture surface is observed in the case of
    a) Shear failure
    b) cup and code failure
    c) Brittle failure
    d) Slant failure
2. The micro-void formation and their coalescence mechanism causes
    a) Ductile failure
    b) Brittle failure
    c) Fatigue failure
    d) All of these
3. The theory of failures suitable for brittle materials is
    a) Maximum distortion energy theory
    b) Maximum shear stress theory
    c) Maximum principal stress theory
    d) Tresca theory
4. The Maximum distortion energy theory holds good for
    a) High brittleness materials
    b) Low ductility materials
    c) High ductility materials
    d) High hardness materials
5. The stress intensity factor of the material, which is independent of the section thickness shows

a) The critical energy release rate
b) Threshold stress intensity factor
c) Two brittle transition parameter
d) Fracture toughness

6. The fracture mechanics-based approach of a design in addition to allowable defect size and applied load uses
   a) Yield strength of the material
   b) Fracture toughness
   c) Ultimate strength of the material
   d) Impact toughness
7. The formation of the plastic zone size ahead of the crack tip depends on
   a) Yield strength of the material
   b) Ultimate strength of the material
   c) Impact toughness
   d) Hot hardness
8. For a material of the given fracture toughness to avoid fracture and increase the allowable crack size
   a) Increases the allowable stress
   b) Decreases the allowable stress
   c) Allowable stress remains the same
   d) All of these are possible
9. If a body is subjected to the stress amplitude lower than the endurance limit then the component
   a) Fractures catastrophically immediately
   b) Fractures after 2×10^6 load cycles
   c) Fractures in the third stage of fatigue
   d) Doesn't fracture
10. The only surface defects are detected using
    a) Dye penetrant test
    b) Magnetic particle test
    c) X ray test
    d) Ultrasonic test
11. A non-destructive technique that is applicable to ferromagnetic materials only is
    a) Dye penetrant test

b) Magnetic particle test

c) X-ray test

d) Ultrasonic test

Answers of Multiple Choice Questions

Key for MCQ: 1 c, 2 a, 3 c, 4 c, 5 d, 6 b, 7 a, 8 b, 9 d, 10 a, 11 b

Short and Long Answer Type Questions

1. What are the different ways the failure of mechanical components can occur?
2. When do we consider that a component has failed?
3. Explain the need for theories of failures for static loading
4. Compare the maximum shear stress theory and maximum distortion energy theory of failure
5. Distinguish the ductile and brittle failures
6. What is the significance of the failure surface in theory of failures?
7. Explain the mechanism of ductile fracture.
8. How does the fracture mechanics approach different from the conventional design approach?
9. What are the different parameters used as a measure of fracture toughness?
10. What is the significance of fracture toughness?
11. How does the fracture toughness differ from the impact toughness?
12. Establish the equation for the stress required for brittle fractures using Griffith theory of failure.
13. Explain the underlying principle of various theories of failure under static loading for ductile and brittle materials.
14. What is fatigue loading, and how is it different from static loading?
15. Explain the approaches used to calculate the various parameters of the fatigue loading like mean stress and stress amplitude, stress range, and maximum stress.
16. Explain the SN curve and its significance in respect to component design and fatigue loading.
17. How to determine the fatigue strength of material using the SN curve?
18. What is the concept of the endurance limit and the fatigue life of the component?
19. How does the crack propagate during the fatigue as a function of a number of load cycles?

20. Explain the stable crack growth as a function of stress-intensity-factor range using suitable schematic diagrams and power law.
21. Explain the different stages of fatigue fractures.
22. How do the metal properties affect the different stages of fatigue fracture?
23. Explain the cumulative fatigue damage approach to determine the fatigue life.
24. What is the importance of non-destructive techniques in the manufacturing and service of mechanical components?
25. Explain the methodology used to conduct the dye penetrant and magnetic particle test.
26. What is the underlying principle of ultrasonic and x-ray testing?

Numerical Problems

1. Principal stresses at a point in a component are 100 MPa and -40 MPa respectively. Considering permissible stress of 300 MPa as per simple tensile and compressive test, calculate the factor of safety (FOS) using given data for each of the following theories of failure: Consider Poisson's ratio (μ) = 0.3
    a) Maximum principal stress theory
    b) Maximum shear stress theory
    c) Maximum distortion energy theory
2. A cube of 10 mm side is subjected to tensile loading in x and y directions of 3000 N, 2000 N respectively. Additionally, a shear load of 500 N is acting on the x-y plane. The component has a yield strength of 30 MPa. Determine the following according to Von-Mises theory.
    a) Principal stresses
    b) Failure possibility without considering FOS
    c) Factor of safety
3. The tensile and shear loads acting on a steel rivet are 40 kN and 20 kN respectively. Considering the yield strength of 200 MPa, calculate the rivet diameter using given data for following theories of failure.

    Factor of safety = 2, Poisson ratio (μ) = 0.3

    a) Maximum principal stress theory
    b) Maximum shear stress theory
    c) Maximum distortion energy theory
4. The yield strength of a steel is found to be 400 MPa in a simple tensile test. According to maximum principle stress theory, calculate the working stress in steel component for a given factor of safety 2.

5. A stress element is subjected tension and shearing stress of 60 MPa and 40 MPa respectively. Considering the yield point of 200 MPa in a component of the ductile metal during tension, calculate the factor of safety based on maximum shear stress theory.
6. The yield and ultimate strength of a carbon steel is 450 and 600 MPa respectively. The endurance limit (stress amplitude) of the steel obtained rotatory fatigue testing is 180 MPa. If a component is to be designed for a fatigue load condition with a mean stress of 100 MPa. Determine the fatigue strength (stress amplitude) as per Soderberg, Gerber and Goodman approaches.
7. The fracture toughness of very wide high strength steel plate is 200 $MPa.m^{1/2}$. The tensile stress applied on the plate is 200 MPa. Determine the critical crack size.

PRACTICAL

1. Conduct fracture toughness test of high strength steel using CT specimen. Plot load-extension curve and calculate fracture toughness.
2. Perform the Dye penetrant test to detect the surface defect and report the observation
3. Conduct the magnetic particle test on weld joint of cast iron report the observations
4. Perform rotary fatigue testing using varying loads to develop S-N curve for carbon steel and establish the endurance limit.

KNOW MORE

Observe the failed components of ductile and brittle materials and make macro-observation of the fracture surface, cross-sectional variation of fractured component, possible crack nucleation and growth direction. Mechanical components like shaft, axle, and chassis subjected to fatigue, and make macro-observations on the fracture surface, cross-section of fractured component, identify the zones of different stages of fatigue. Explore the newer non-destructive testing methods like optical, thermal imaging, ultrasonic array based techniques.

REFERENCES AND SUGGESTED READINGS

Suggested resources for further reading/learning

1. G H Ryder, Strength of materials, ELBS, McMillan (1988)
2. GE Dieter, Mechanical metallurgy, McGraw Hill, Publication (1989)
3. D K Dwivedi, Fundamentals of Metal Joining, Springer nature, Singapore (2022)

4. Chaitanya Sharma, Dheerendra Kumar Dwivedi, Pradeep Kumar, Fatigue behavior of friction stir weld joints of Al–Zn–Mg alloy AA7039 developed using base metal in different temper condition, Materials & Design, Volume 64, December 2014, Pages 334-344
5. Ratna Kumar, NPTEL Course "Basics of Materials Engineering" https://onlinecourses.nptel.ac.in/noc20_me78/preview
6. https://www.youtube.com/watch?v=_DH3546mSCM
7. https://mechanicalc.com/calculators/fracture-mechanics/

Dynamic QR Code for the Unit 3

Unit 4

Phase Diagram and Metallurgical Transformation

Alloys, Phases, Phase diagram, Microstructure development, Iron-Carbon Diagram

Unit Specific Learning Objective

The objective of this unit is to talk about the following aspects

- To understand the effect of the temperature on phase transformation
- To know the phase transformations occurring during heating and cooling in binary alloys
- To estimate the number of phases in a binary alloy as a function of the temperature, pressure and composition
- To determine the proportion and composition of phases at a temperature of interest using binary phase diagram
- To learn about the lever rule and its application in phase diagram
- To understand the various phase transformations in Fe-C system using iron-carbon phase diagram
- To learn about the classification of carbon steel and cast iron based on composition

Additionally, few fundamental questions for self-assessment based on fundamentals **are** included in this chapter in form of recall, application, comprehension, analysis and synthesis. There are further suggested readings and references for deep learners and reader's assistance.

Rationale

The metals during manufacturing and service are subjected to various thermal cycles. Thermal cycles involving heating, soaking at high temperatures, and cooling at different rates cause many metallurgical changes. These metallurgical transformations affect the microstructure (phase and grain structure). A metal's microstructure significantly determines the mechanical, tribological and corrosion properties. Therefore, it is important to understand the various changes in phase and grain structure taking place in a metal due to thermal cycling imposed during the manufacturing and service. The study of phase diagrams of binary alloys significantly help in understanding and estimating the phase transformation and microstructural changes which can occur due to the thermal cycle imposed. This unit presents the phase rule, phase diagrams of the different types of binary alloys, iron carbon diagram, various isothermal transformations observed in Fe-C in systems.

Pre-Requisites

A course on chemistry: Class XII

Learning outcomes

U4-O1: Ability to estimate the number of phases at a given temperature in a binary alloy

U4-O2: Ability to estimate and predict the phase transformations and microstructure of binary alloy subjected to the thermal cycles

U4-O3: Ability to estimate the temperature at which phase transformations besides estimation of the compositions and proportions of various phases at different temperatures in binary alloys

U4-O4: Ability to understand the phase transformation of iron-carbon system during heating and cooling and relate the microstructure of Fe-C system with the mechanical properties

U4-O5: Ability to design the microstructure of binary alloys to achieve the desired Mechanical properties

Unit-4 Outcomes	*EXPECTED MAPPING WITH COURSE OUTCOMES (1- Weak Correlation; 2- Medium correlation; 3- Strong Correlation)*					
	CO-1	CO-2	CO-3	CO-4	CO-5	CO-6
U4-O1	1	3	3			
U4-O2	1	3	3			
U4-O3	1	3	3			
U4-O4	1	3	3			
U4-O5	1	3	3			

Course Objective

1. The student will be able to identify crystal structures for various materials, and one could understand the defects in such structures.
2. Understand how to tailor the material properties of ferrous and non-ferrous alloys.
3. How to quantify mechanical integrity and failure in materials?

4.1 Introduction

The microstructure of metals significantly determines the metals' mechanical and tribological properties. The microstructure of metal shows the grain and phase structure. The phase structure of metal exhibits the type of phases, the relative amount of the various phases, and their distribution, while the grain structure shows the size, shape, and distribution of the grains (micro-constituents). The microstructure of a metal depends on the chemical composition, cooling conditions during the solidification, and mechanical and thermal treatments if any. Metal can be pure, alloy, compound, and composite. This unit primarily focuses on understanding the metallurgical transformations in metals with the help of phase diagrams. Additionally, the microstructure development through various (monotectic, eutectic, peritectic, eutectoid) metallurgical reactions observed in phase diagrams helps in estimating and predicting microstructures of metal processed through the casting and heat treatment. Steel, cast iron, and wrought iron are the most commonly used ferrous metals. Common metallurgical transformations occurring in the Fe-C system can be explained with the help of an iron-carbon phase diagram.

4.2 Alloy

An alloy is a mixture of two or more elements, out of which at least one element is a metal. The metal matrix acts as a solvent while other elements (metallic/non-metallic) act as solutes that get dissolved in the metal matrix to form an alloy. The solvent is a major component, while the solute is a minor component in a solid solution forming an alloy. However, there is a solubility aspect also in solid solutions. Solubility indicates the amount of solute which can be dissolved in a solvent. The solubility, in general, increases the temperature. The variation in solubility as a function of temperature forms a basis of precipitation hardening.

Depending upon the number of components (elements) present in an alloy, these are termed binary, tertiary, and quaternary alloys. A binary alloy is composed of two components, a tertiary alloy has three components, and a quaternary contains four.

Atoms of an alloying element get dissolved in the metal matrix to form a solid solution primarily in two ways: substitutional solid solution and interstitial solid solution. A substitutional solid solution is developed when atoms of alloying elements replace the atoms of solvent from the regular crystalline structure. On the other hand, an interstitial solid solution is formed when atoms of alloying elements are accommodated in the vacant spaces between the atoms of a regular crystalline structure. The presence of an alloying element in a metal Matrix as a solute increases the yield strength. However, the extent of increase in strength with the addition of alloying elements in pure metals depends upon the difference in the atomic size of solute and solvent and the concentration (wt. %) of the alloying elements. In general, an increase in the difference in the atomic size of solvent and solute and the concentration of the alloying elements in the metal matrix increases the yield strength. Further details can be seen from Unit 1 under the metal strengthening mechanisms.

4.3 Phase Diagram

A phase is a zone in a material with uniform composition, crystal structure, and chemical and physical characteristics. Material may be composed of single or multiple phases. Each phase is separated from other phases by a unique boundary called phase/grain boundary. A single-phase material is called a homogeneous system, and materials with two or more phases are known as a heterogeneous system.

The solubility limit shows how much an alloying element can be dissolved in a phase. There can be various possibilities for solubility, both in the liquid and solid state of a phase. Three scenarios for binary alloys (metallic) are the most common a) completely soluble both in liquid and solid state, b) completely soluble in liquid state but partially soluble in the solid state, and c) completely soluble in liquid state but insoluble in a solid state. In this unit, phase diagrams of binary alloys only will be presented.

4.3.1 Gibbs Phase rule

The phase diagram show phases in equilibrium for a given set of temperature, pressure and composition. A system is considered in equilibrium if, at a given temperature, pressure and composition, it remains stable as a function of time. These three conditions (temperature, pressure and composition) significantly determine the phase transformations occurring in a phase diagram. For example, in one component system, water is found in three different states at varying temperatures and pressure through different processes (Fig. 4.1).

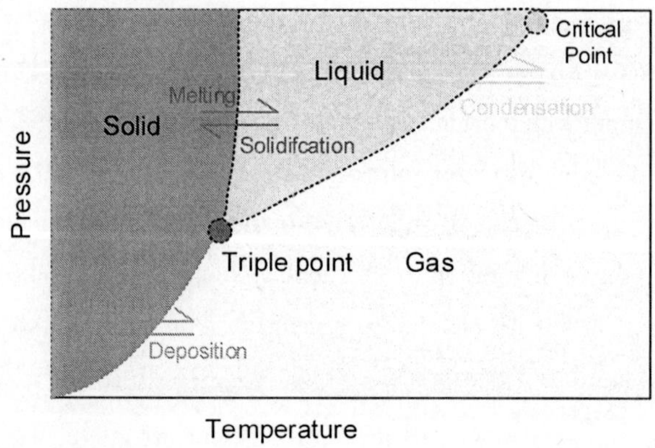

Figure 4.1: Schematic diagram showing changes in states of water as a function of temperature and pressure

The phase diagram shows the temperature at which transformations take place, the number of phases present at a given temperature in given alloy composition, and the variation in phases with alloy compositions. The relationship between the number of phases, the number of components in the alloy (say 2 in binary alloy), and the degree of freedom/number of variables is expressed using the Gibbs phase rule.

$$F= P + C - 2$$

Where F is the degree of freedom (temperature, pressure and composition), P is the number of phases, and C is the number of components (elements) in an alloy. Processes like casting, welding, and forming heat treatment of metallic alloys are done under atmospheric conditions. In the case of transformations (in the processing of metallic alloys) taking place at atmospheric pressure, the above equation is rewritten as:

$$F= P + C - 1$$

A binary alloy phase diagram is developed using cooling curves obtained during the solidification of many-many binary alloys with gradually changing wt. % of components A & B (100% A, 80%-20%B, 60%A-40%B, 50%A-50%B, 40%A-60%B, 20%A-80%B, 100%B). Information extracted from the cooling curves in the form of start, and end temperature of phase transformation, like liquid to solid during solidification, is obtained. Pure metals solidify at a particular (constant) temperature (Fig. 4.2a). The solidification starts with nucleation, as indicated by undercooling, followed by growth at a constant temperature. The degree of undercooling (3-10° C) indicates resistance to nucleation.

Alloys on the other hand, solidify over a range of temperatures with two different solidification start and end temperatures (Fig. 4.2b). The difference between solidification start and end temperatures is commonly termed as solidification temperature range. A line connecting the solidification start temperatures of binary alloys with varying compositions appears as liquidus while a line connecting solidification end temperatures of binary alloys with varying compositions is shown by solidus in phase diagrams.

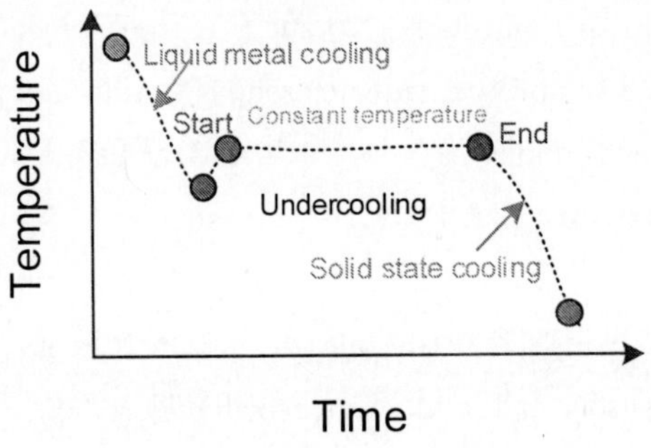

a)

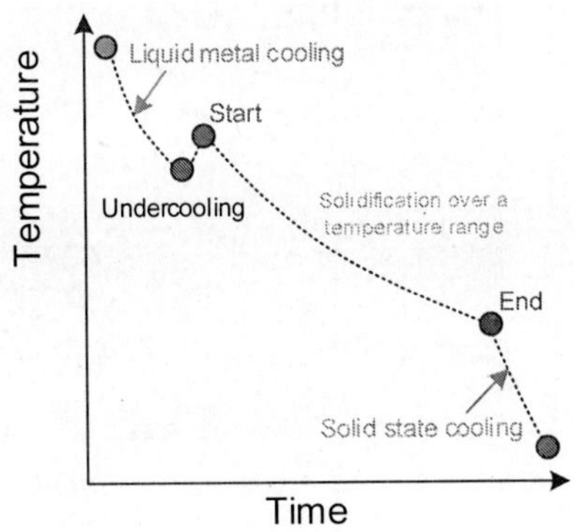

b)

Figure 4.2: Schematic diagram showing cooling curves during solidification of a) pure metals and b) alloy other than eutectic composition

4.3.2 Binary Isomorphous Metal System (Complete liquid and solid state solubility)

The binary alloys have two components if they have complete solubility in liquid and solid state called isomorphous system. Unlike pure metal, such a system shows a range of temperatures over which solidification takes place instead of a single and constant temperature.

Binary alloys show complete solid and liquid state solid solubility when both elements of the alloy have similar a) atomic radii, b) crystal structure, c) electronegativity (else reaction may lead to compound formation). Additionally, the solute element should have higher valence than the solvent.

The binary alloy of two elements A (Ni) and B (Cu) of a metal system having complete solid and liquid state solubility. An isomorphous equilibrium phase diagram consists of two lines (liquidus and solidus) only at which phase transformation is observed (Fig. 4.3). One line corresponds to solidus (lower), and another is liquidus (upper). The region between these two lines (solidus and liquidus) is called the mushy zone, where both solid and liquid phases co-exist. These two line results in three zones: liquid, mushy (two phases liquid-solid mixture) zone and solid. Above the liquidus, there is a single-phase zone (only liquid); similarly, below the solidus, there is a single-phase region (only solid) as both elements have complete solid and liquid state solubility.

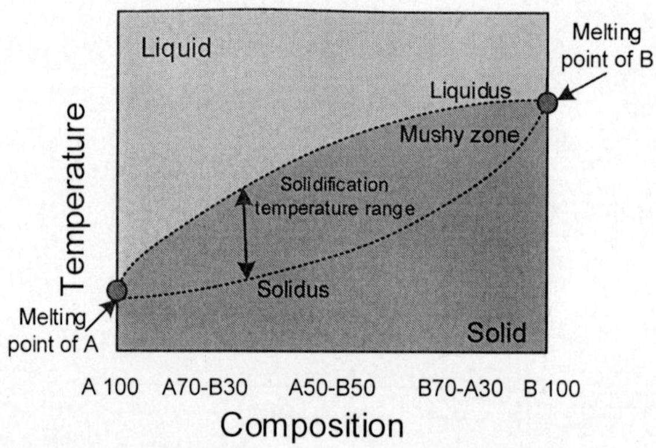

Figure 4.3: Schematic diagram showing binary phase diagram of the isomorphous metal system

From any phase diagram following information can be obtained:

- Variation in phases with the temperature and alloy compositions.
- The temperature of various phase transformations in a given alloy system.
- Amount of phases present at any temperature for a given alloy composition.
- Variation in composition of various phases with change in temperature.

- Variation in solubility of alloying elements in a solvent.

Line "PQ" shows the composition of an alloy in the equilibrium diagram is called composition line (Fig. 4.4). With the decrease in temperature of the molten metal from liquid state "R" no phase change takes place until it crosses liquidus at point "S". Intersection point of liquidus at "S" by composition line "PQ" indicates the temperature at which solidification starts and the intersection of composition line with solidus at "T" shows completion of solidification. A horizontal line parallel to the X-axis passing through the temperature of interest (any temperature on the ordinate) is called tie line "U-V-W". Projection of intersecting points of tie line "U-V-W" with liquidus and solidus on X axis indicates the composition of liquid and solid phases present at that temperature. Consider a tie line length "U-W" and length of tie line in numbers can be obtained from the difference in wt.% of any one component (A, B) say "Y-X".

Binary alloy composition point on the tie line (V) is called fulcrum. Length of tie-line on liquidus side (UV) and that on solidus side (VW) with respected to length of total tie line (UW) indicates the proportions of solid (UV/UW) and liquid (VW/UW) phases present in mushy zone. This is called lever rule and can be used to find the composition and proportion of phases present in equilibrium diagram at any temperature.

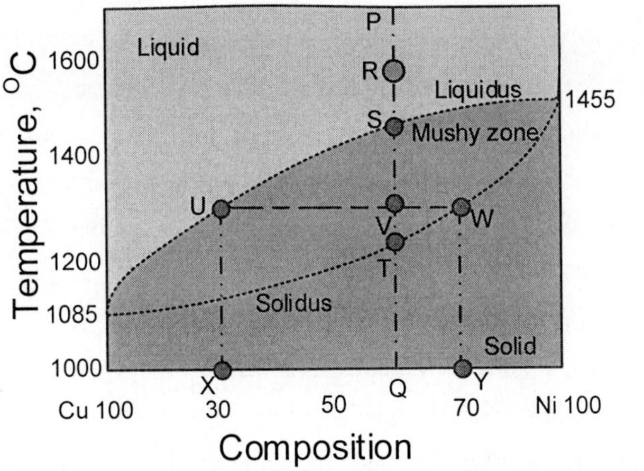

Figure 4.4: Binary phase diagram of isomorphous Ni-Cu metal system

Number of phases: Number of phases co-existing at any point (i.e. the combination of temperature or composition) in the phase diagram can be obtained by simply drawing a tie at that point and noting how many phase boundaries are cut by tie line. Number of phases in any area of phase diagram becomes equal to number of phase boundaries cut by tie line at given temperature.

Composition of phases: Composition of terminal phases can be obtained by projecting the point of intersection of tie line and phase boundary on X axis. Location of projection on X axis gives the composition of co-existing phases.

Relative amount of phases: Draw a horizontal line at the temperature of interest parallel to X axis and locate two points of intersection of tie line with phase boundaries, which will determine the length of tie-line. Point of intersection of tie line with composition line is termed fulcrum of lever (whose length is equal to tie line length). Length of tie line from the fulcrum to the one end becomes proportional to the amount of the phase in opposite side (Fig. 4). For example, proportion of liquid phase will be equal to the (VW/UW) while the proportion of solid phase will be equal to the (1- VW/UW). This is commonly known as lever rule. It can be applied in any portion of the equilibrium phase diagram with any type equilibrium diagram.

4.3.3 Completely soluble in liquid state and insoluble in solid state

Alloy systems like lead-tin having complete solubility in liquid state but insoluble in solid-state show completely different behaviour (than the isomorphous systems) on cooling from molten state to the room temperature under equilibrium conditions (Fig. 4.5). To understand the construction of this phase diagram it is necessary to know the Roult's law. This law states that addition of an alloying element which is insoluble in solid state in pure metal reduces the melting point of solvent. Addition of such as alloying element in a metal matrix reduces the melting point. This reduction in melting temperature of the alloy continues with the increase in percentage of alloying element up to a fixed composition. Thereafter, rise in melting point of the alloy starts. That is why only for one fixed composition such as alloy melts/solidifies at the lowest temperature. The composition, at which such as alloy melts at lowest temperature is called eutectic composition and corresponding temperature, is called eutectic temperature. The point in phase diagram indicating the eutectic composition and eutectic temperature is called eutectic point. Liquid to solid state transformation at the eutectic point takes place at constant temperature whereas transformation other than eutectic composition takes place over a range of temperature. In general, all those liquid to solid state transformations which take place over a range of temperature increase the tendency of hot cracking and gaseous defects.

While pure metals and eutectic alloys, which solidify at constant temperature show higher tendency of the sound and defect free castings.

Alloy systems having concentration of the alloying element less than that is required to form eutectic are called hypoeutectic alloy and those having alloying element more than that is required for eutectic composition are called hypereutectic alloys. For example, in Fe-C alloy system, eutectic is formed with 4.3% C whereas in Al-Si alloy eutectic is formed with 12.3% silicon.

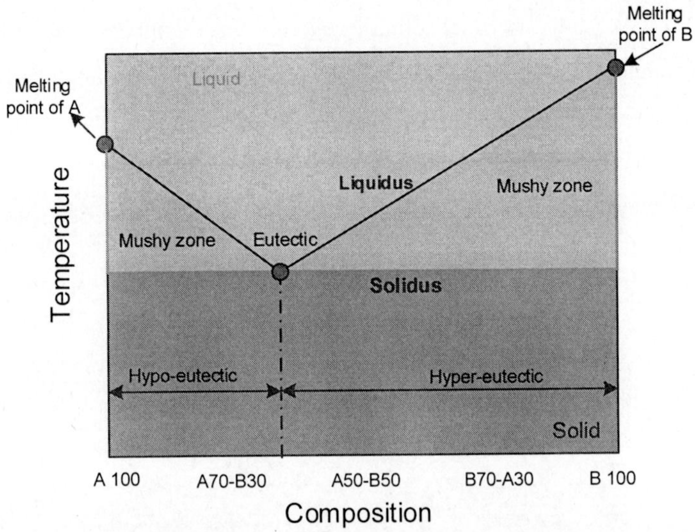

Figure 4.5: Binary phase diagram of alloy system having complete solubility in liquid state and insoluble in solid state

Hypoeutectic alloy (completely soluble in liquid state and insoluble in solid state) (say 20%B and 80%A) during cooling from point 'B' does not show any phase transformation until temperature crosses the liquidus line at point 'C' (Fig. 4.6). At temperature corresponding to point 'C' solidification of element A starts. The composition of solidifying constituent can be checked by drawing a tie line like “H-C” at any temperature of interest (between liquid and solidus) parallel to the X axis and obtaining the projection of point ‘H’ on the X axis. The projection of ‘H’ on the X axis indicates the composition of element A. Since both elements are insoluble in solid state therefore solidification of such an alloy results in pure grains of A and B. Further, decreases in temperature increase the percentage of the solid phase, i.e. A. Composition of liquid changes continuously with a reduction in temperature, and it moves along the line C-G. Let's consider a tie line "E-F-G" at a little lower temperature wherein F corresponds to composition point "fulcrum". The tie line intersects solidus and liquidus at "E" and "G" point. The project of points "E" and "G" point on X-axis will show the composition of solid and liquid metal, respectively, present at this temperature. Similarly, the ratio of EF/EG and FG/EG will indicate the proportion of liquid and solid phases, respectively, in the mushy zone at this temperature. Careful observation reveals that the liquid metal is progressively enriched with the element B. As the temperature of liquid metal reaches the point 'I,' i.e. corresponding to the eutectic one and rest of the liquid metal solidifies by eutectic reaction, wherein a liquid metal of eutectic composition solidifies at constant temperature into a fine mixture of two solids, i.e. A and B phases. The phase solidified before reaching the eutectic point during the solidification is called the pro-eutectic phase (Fig. 4.7). For hypoeutectic alloy, in this case, the pro-eutectic phase is A.

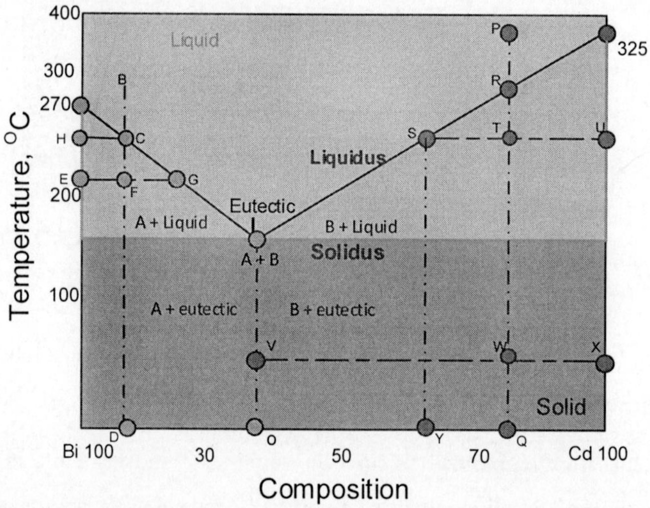

Figure 4.6: Binary phase diagram of Bi-Cd alloy system having complete solubility in the liquid state and insoluble in solid state

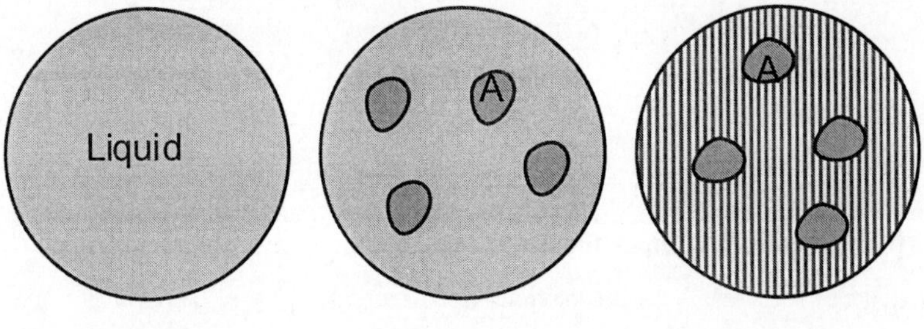

a) b) c)

Figure 4.7: Schematic diagram showing progressive solidification of hypoeutectic alloy a) liquid metal, b) pro-eutectic solid phase and liquid metal, and c) pro-eutectic solid phase and solidified eutectic

An alloy of eutectic composition does not transform until liquid metal attains eutectic temperature (T_I). At eutectic temperature (T_I), liquid metal transforms into two types of solid phases, i.e. A and B. A and B crystallize in pure form as they are insoluble in the solid state. Each phase precipitates alternatively, resulting in the lamella structure of eutectic phases (Fig. 4.8). Growth of each phase depends on the rate of cooling, i.e. the time available for growth of each phase. Transformation of one phase from liquid metal (say A) into solid depletes the liquid of element A; therefore, the concentration of element B in liquid metal increases. Hence, the crystallization of element B from the liquid metal starts. Crystallization of B depletes the liquid metal of B element and thereby enriching the liquid metal with element A. This way, alternate A and B precipitation from liquid metal into solid phases continue until the whole liquid metal solidifies. It means that the composition point of liquid metal during the transformation moves (left & right side) about the eutectic point instead of being fixed.

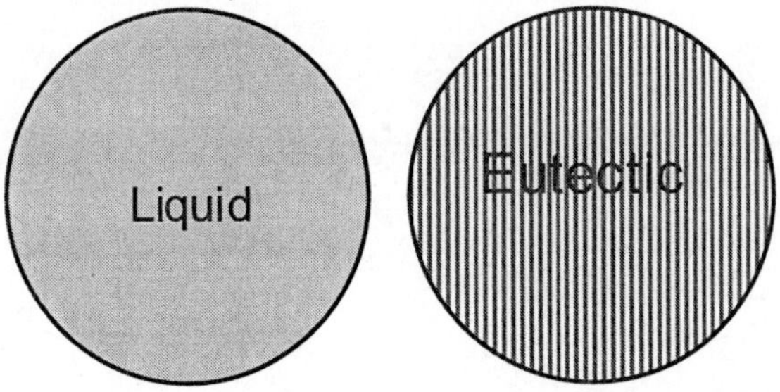

a) b)

Figure 4.8: Schematic diagram showing solidification of eutectic alloy a) liquid metal, and b) solidified eutectic

Alloys of hypereutectic composition solidify in the same way as hypoeutectic alloys, except that instead of element A as the primary (pro-eutectic) phase, element B appears as the primary phase and subsequently, eutectic transformation takes place in the same fashion. Similarly, transformation with the reduction in temperature, the composition of phases during solidification and their relative amount can also be obtained as described above for hypoeutectic alloy.

Hypoeutectic alloy (completely soluble in liquid state and insoluble in solid state) (say 80%B and 20%A as shown by composition line PQ) during cooling from point 'P' does not show any phase transformation until temperature crosses the liquidus line at point 'R' (Fig. 4.9). At temperature corresponding to point 'R' solidification of element B starts. The composition of solidifying constituent can be checked by drawing a tie line like "R-Z" parallel to the X axis and obtain the projection of point Z' on X axis. Projection of 'Z' on X-axis indicates the composition of element B as both elements are insoluble in the solid state. Further, decrease in temperature increases the percentage of the solid phase, i.e. B. Composition of liquid changes continuously with the reduction in temperature and moves along the line R-S.

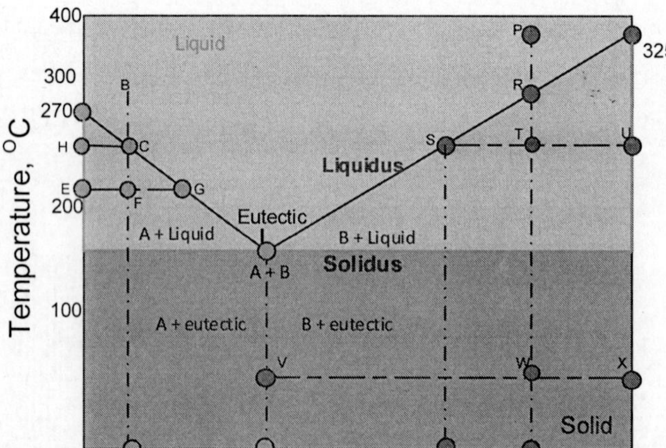

Figure 4.9: Binary phase diagram of Bi-Cd alloy system having complete solubility in liquid state and insoluble in solid state

Let's consider a tie line "S-T-U" at a slightly lower temperature wherein T corresponds to the composition point "fulcrum". The tie line intersects solidus and liquidus at "U" and "S" point. The projection of points "U" and "S" point on the X axis will give the composition of solid and liquid metal, respectively, present at this temperature. Similarly, the ratio of TU/SU and ST/SU will indicate the proportion of liquid and solid phases in the mushy zone at this temperature. Careful observation reveals that the liquid metal is progressively enriched with element A, and the concentration of element B decreases continuously. As the temperature and composition of liquid metal reach the point 'I' i.e. corresponding to the eutectic point, the rest of the liquid metal solidifies by eutectic reaction where a liquid metal of eutectic composition solidifies at constant temperature into a fine mixture of two solids i.e. A and B phases. The phase solidified before reaching the eutectic point during the solidification is called the pro-eutectic phase (Fig. 4.10). For hypereutectic alloy in this case, the pro-eutectic phase is B.

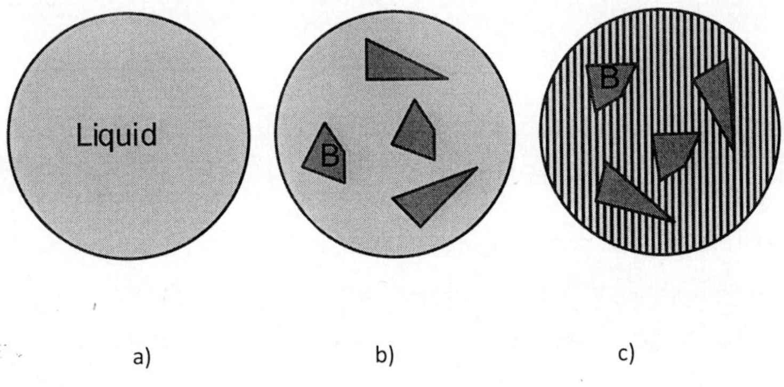

Figure 4.10: Schematic diagram showing progressive solidification of hypereutectic alloy a) liquid metal, b) pro-eutectic solid phase and liquid metal and c) pro-eutectic solid phase and solidified eutectic

4.3.4 Completely soluble in the liquid state and partially soluble in solid state

In this case, elements of an alloy system have partial solubility in the solid state in addition to complete solubility in the liquid state of the alloy. In most of the alloy systems, this kind of transformation is observed as most of the alloying elements have some solid solubility in other elements as solvent. However, the limit of solubility may vary from very small (0.002%) to 99 or 100 %. This is the most common and important type of alloy system like Fe-C, Al-Si, Al-Cu etc. The metallurgical transformation occurring in alloy systems with partial solid-state solubility forms the basis for various heat treatments (precipitation hardening, annealing, normalizing, hardening) to realize the desired properties in metal by controlling the microstructure.

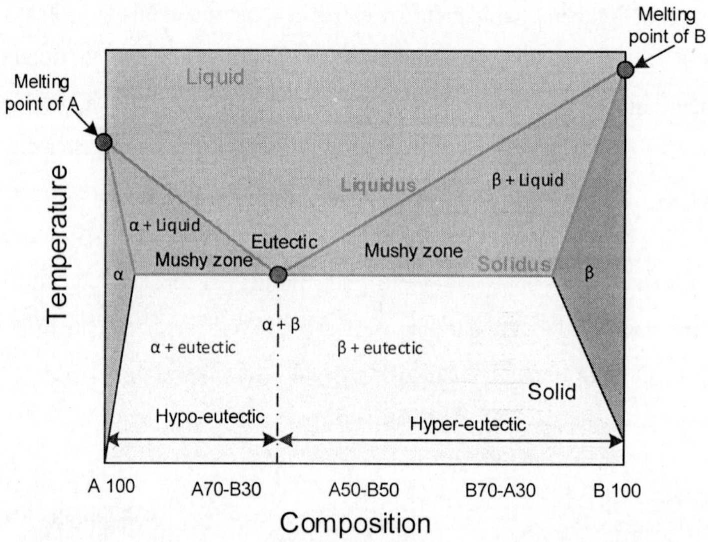

Figure 4.11: Binary phase diagram of alloy system (A-B) having complete solubility in liquid state and partial solubility in solid state

Broadly construction of the phase diagram for this type of alloy system is similar to that of the earlier one (completely soluble in the liquid state and insoluble in solid state) except that instead of pure elements (due to complete insolubility in solid state) as terminal phases; solid solutions (α, β) of elements are formed (Fig. 4.11). For example, two elements A & B form a solid solution. If B is solute then the solid solution formed is "α", and if A is the solute, then the solid solution formed is β. The composition of terminal phases (α, β), depends upon the solubility limit at room temperature. In addition to the above, there is solvus lines for both elements of the alloys which indicate solid solubility limit (for A in B or B in A) as a function of temperature. The solvus line shows the decrease in solubility limit with the reduction in temperature. The concept forms the basis of precipitation hardening heat treatment of a few alloys.

Hypoeutectic alloy (completely soluble in a liquid state and partially soluble in a solid state) (say 70%B and 30%A as shown by composition line PQ) during cooling from point 'P' does not show any phase transformation until temperature crosses the liquidus line at point 'R' (Fig. 4.12). At temperature corresponding to point 'R' solidification of phase β starts. The composition of solidifying constituent β can be checked by drawing a tie line like "R-Z" parallel to the X axis and obtaining the projection of point Z' on X axis. The projection of 'Z' on X-axis indicates the composition of phase β. Further, decreases in temperature increase the percentage of the solid phase, i.e. β. The composition of liquid changes continuously with the reduction in temperature and it moves along the line R-S.

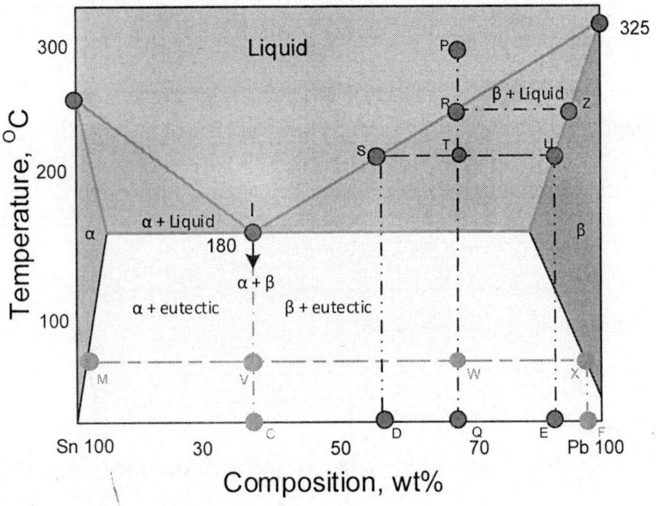

Figure 4.12: Binary phase diagram of Pb-Sn alloy system having complete solubility in the liquid state and partial solubility in solid state

Let's consider a tie line "S-T-U" at a slightly lower temperature wherein T corresponds to the composition point "fulcrum". The tie line intersects solidus and liquidus at "U" and "S" point. The projection of points "U" and "S" point on the X axis will give the composition of solid phase "β" and liquid metal, respectively, present at this temperature. Similarly, the ratio of TU/SU and ST/SU will indicate the proportion of liquid and solid phase "β" respectively, in the mushy zone at this temperature. Careful observation reveals that the liquid metal is progressively enriched with element A, and the concentration of element B decreases continuously. As the temperature and composition of liquid metal reach the point 'I,' i.e. corresponding to the eutectic point and rest of the liquid metal solidifies by eutectic reaction where a liquid metal of eutectic composition solidifies at constant temperature into a fine mixture of two solid (α, β) phases. The phase solidified before reaching the eutectic point during the solidification is called the pro-eutectic phase "β". For hypoeutectic alloy pro-eutectic phase will be "α".

Similarly, the proportions of phases in the solid state of hypereutectic alloy (as per composition line "PQ") at a temperature corresponding to the tie-line "M-W-X". The tie line intersects two phase boundaries: α on the left and β on the right. For a given alloy composition at fulcrum "W", the composition of α and β in the alloy can be obtained using projections from M and X point on the X axis, while and proportion of α and β in the alloy can be obtained using lever rule. , The proportion of α and β can be expressed using XW/MX and MW/MX, respectively.

Since the alloy is in consideration in hypereutectic forming pro-eutectic phase "β" and eutectic mixture "α-β". For a given alloy composition at fulcrum "W", the composition of pro-eutectic phase β and eutectic mixture in the alloy can be obtained using projections from V and X point on the X axis as the tie line is revised and reduced to "V-W-X". The proportion of pro-eutectic phase β and eutectic in the alloy can be obtained using the lever rule, i.e. VW/VX and WX/VX, respectively.

4.4 Common Reactions in Phase diagrams

4.4.1 Peritectic reaction

The peritectic reaction occurs at specific composition and temperature of binary alloy wherein one solid phase and liquid metal transforms into another type of solid (Fig. 4.13). For example, Fe-C system shows the peritectic reaction at 0.18%C and 1495 °C wherein δ-Fe and liquid metal transforms into γ-Fe.

Solid phase 1 (α) + *Liquid metal* ⇌ *Solid phase 2 (β)*

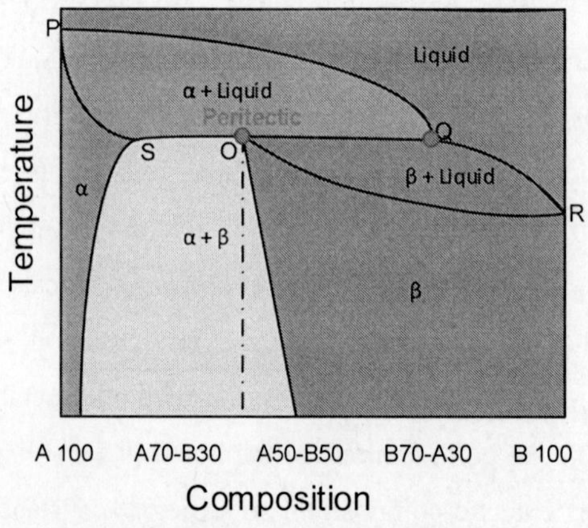

Figure 4.13: Binary phase diagram of alloy system (A-B) having complete solubility in the liquid state and partial solubility in solid state subjected to "peritectic reaction"

4.4.2 Monotectic reaction

Few binary alloys show complete insolubility in the liquid state (over a certain range of compositions) and partial solubility in the solid state. Therefore, the two elements of the alloy remain separate and do not mix (L1 and L2). The monotectic reaction occurs at specific composition and temperature (say reaction point R) or over a range of composition and temperature of binary alloy (S-T) wherein one liquid phase metal transforms into a solid and another type of liquid (Fig. 4.14). For example, Fe-C system shows the peritectic reaction at 0.18%C and 1495 °C wherein δ-Fe and liquid metal transforms into γ-Fe.

Liquid metal 1 $\rightleftharpoons$ *Solid phase 1 (α)* *+ Liquid metal 2*

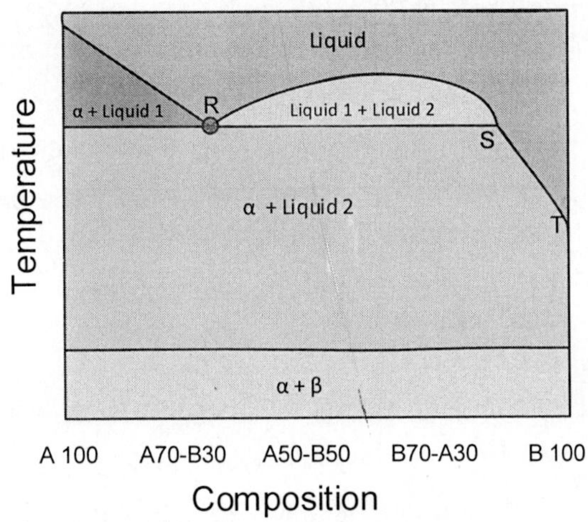

Figure 4.14: Binary phase diagram of alloy system (A-B) having limited or no miscibility in the liquid state and insoluble in solid state subjected to "monotectic reaction"

4.4.3 Peritectoid Reactions

The peritectoid reaction is a solid state transformation reaction observed in a few binary alloys. In a peritectoid reaction, two solid phases (β, γ) transform into a new solid phase (α) as shown in Fig. 4.15. Cu-Sn phase diagram shows several peritectoid reactions between 32 and 40% wt.% Sn where a peritectoid reaction between the γ gamma solid solution and the intermetallic compound Cu_3Sn forms Cu_4Sn.

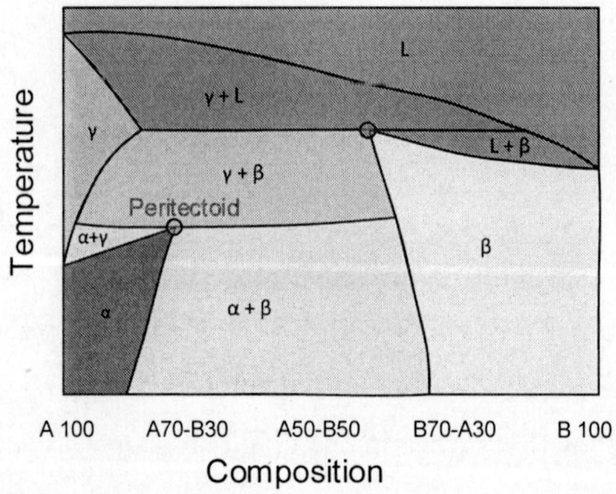

Figure 4.15: Binary phase diagram of alloy system (A-B) having solubility in the liquid state and partial solubility in solid state subjected to "peritectoid reaction"

4.5 Fe-C Equilibrium Phase Diagram

Fe-C phase diagram is an example of complete liquid state solubility and partial solubility of alloying element "C" in solid state. Fe-C diagram is also called iron-iron carbide diagram because these are the two main phases observed at room temperature. In contrast, the presence of other phases depends on temperature, type and amount of alloying elements present in the iron.

4.5.1 Allotropy and critical temperatures

Change in the crystal structure of an element with the rise in temperature is termed allotropy. Iron shows the allotropic behaviour at temperatures 910 and 1390 °C. Iron changes its crystal structure first from B.C.C to F.C.C. at 910 °C and then from F.C.C to B.C.C. at 1395 °C (Fig. 4.16). Therefore, solubility of carbon in iron changes significantly with the rise in temperature especially above 910 and 1395 °C.

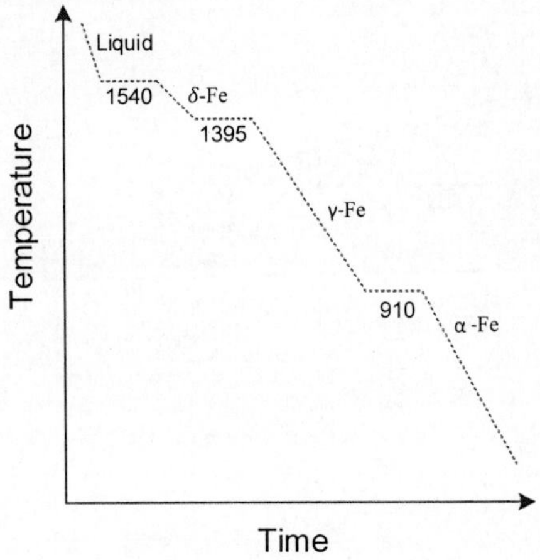

Figure 4.16: Schematic diagram showing allotropic behaviour Fe as a function of temperature

Fe-C shows the various reactions and phase transformations, which occurs on cooling Fe-C, alloy system from molten state to the room temperature under equilibrium condition or heating from room temperature to molten condition (Fig. 4.17). There are three most common reaction points observed in Fe-C diagram namely peritectic reaction, eutectic reaction and eutectoid reaction. It is normally plotted for carbon up to 6.67%, as iron with more than 6.67% carbon is rarely used due to inferior mechanical properties.

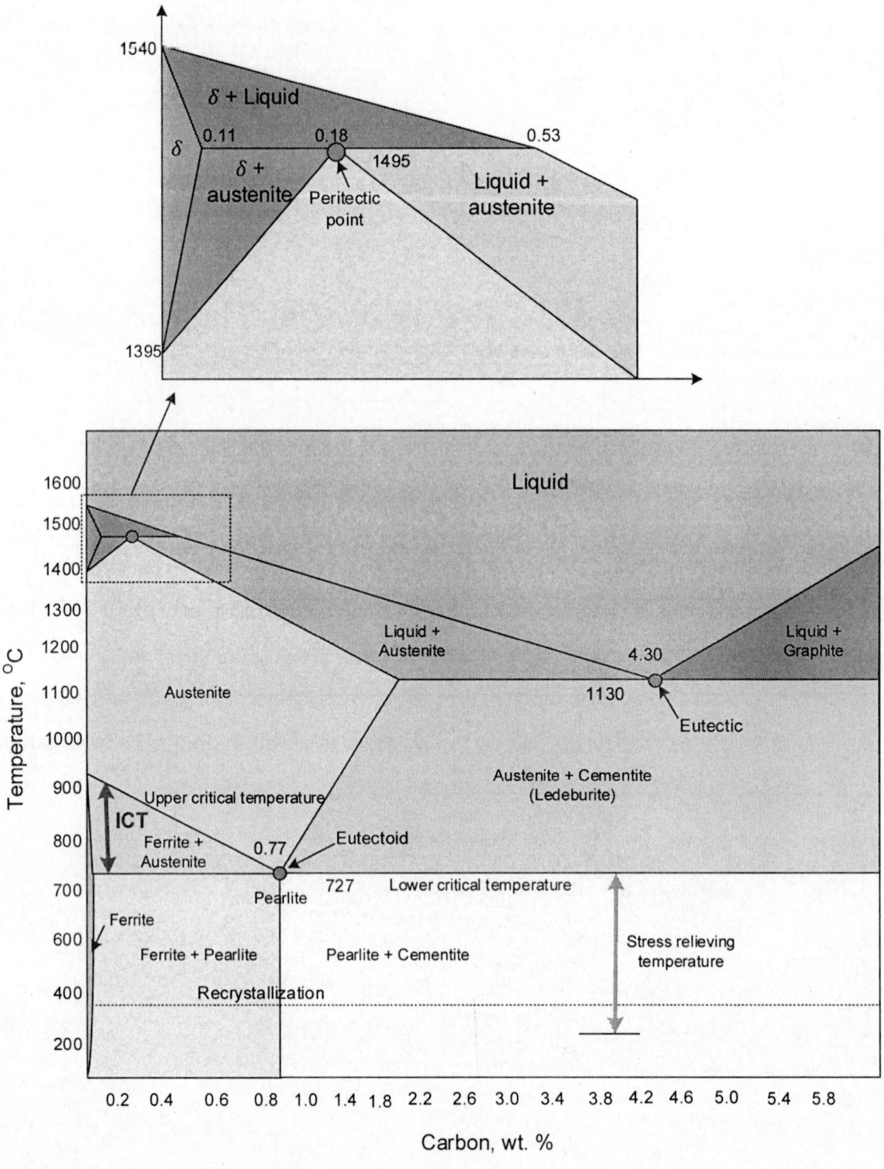

Figure 4.17: Fe-C phase diagram

4.5.2 Isothermal Transformations in Fe-C diagram

Metallurgical transformation occurring in Fe-C system during cooling from a molten state to room temperature or heating from room temperature to molten state under equilibrium conditions can be understood and analysed by drawing a composition line and observing its intersection with the phase boundaries. There are three isothermal reaction points in the Fe-C diagram, namely peritectic, eutectoid and eutectic. All three transformations at these reaction points occur at a fixed temperature and composition as mentioned below.

a) *Peritecitc reaction takes place at 1495 ^{0}C and 0.18% C*

$$\textit{Liquid metal} + \delta\textit{-iron} \rightleftharpoons \gamma\textit{-iron}$$

b) *Eutectic reaction occurs at 1130 ^{0}C and 4.3% C*

$$\textit{Liquid metal} \rightleftharpoons \gamma\textit{-iron} + Fe_3C$$

c) *Eutectoid (solid state) reaction occurs at 727 ^{0}C and 0.8% C*

$$\gamma\textit{-iron} \rightleftharpoons \alpha\textit{-iron} + Fe_3C$$

Proportions of various phases of these transformations can be obtained using the lever rule. Fulcrum of the lever depends upon the alloy composition, i.e. carbon content. Since these transformations occur at constant temperatures, it is easy to find a tie line. Terminal phases (and their compositions) can be obtained using the tie line and alloy compositions.

Further, the phases appearing in Fe-C diagram are ferrite (α-Fe), austenite (γ-Fe), cementite (Fe_3C), δ-iron and mixtures of phases such as pearlite (α-Fe + Fe_3C) and ledeburite (γ-Fe + Fe_3C). Pearlite and ledeburite are the results of eutectoid and eutectic reactions, respectively. Moreover, the ledeburite is a meta-stable phase which eventually appears as pearlite and cementite. The following section discusses these phases and their mechanical properties. The following section presents a few important aspects of these phases and phase mixtures that significantly determine the Fe-C system's mechanical properties.

Ferrite

Ferrite (α-Fe) is an interstitial solid solution of carbon and iron (Fig. 4.18). It has body centred cubic crystal structure. Solubility of carbon in iron at room temperature is negligible (0.005%) while that at eutectoid temperature (727^{0}C) is 0.025%. Ferrite is a soft, low strength, tough and ductile phase.

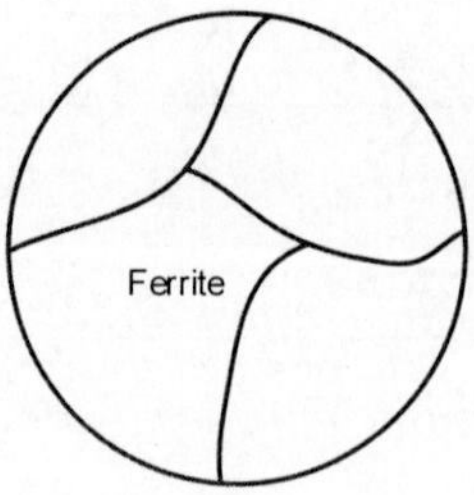

Figure 4.18: Schematic diagram showing optical microscopic features of ferrite (α-Fe)

Austenite

Austenite (γ-Fe) is an interstitial solid solution of carbon in iron. Austenite has face centred cubic crystal structure. This is meta-stable phase in Fe-C diagram which in unstable below eutectoid temperature (727^{0}C). Austenite transforms into pearlite through eutectoid reaction below 727^{0}C. The solubility of carbon in the austenite is a maximum of 2.0% at 1330^{0}C, which reduces to 0.8% with a reduction in temperature to eutectoid temperature (727^{0}C). Austenite can be stabilized at a temperature using suitable alloying (Ni, Mn). Austenite is comparatively harder, stronger, tougher and ductile than the ferrite.

Cementite

Cementite is an inter-metallic compound of iron and carbon, which appears as iron carbide in (Fe_3C) in Fe-C phase diagram. Cementite contains 6.67% of carbon and has an orthorhombic crystal structure. Cementite is the hardest phase appearing in Fe-C diagram while its tensile strength is almost negligible.

Pearlite

Pearlite is a phase mixture of ferrite and cementite. Pearlite is produced by eutectoid transformation (0.8% C, 727 ^{o}C), wherein austenite transforms into pearlite due to the rejection of carbon from austenite beyond the solubility limit below eutectoid temperature. Pearlite has alternate layers (lamellas) of cementite and ferrite (Fig. 4.19). Inter-layer spacing between ferrite and cementite in pearlite is used as a measure of grain size (coarseness/fineness) of the pearlite. A thick layer of ferrite / cementite is considered as a coarse pearlite that is usually produced during annealing heat treatment. The strength of pearlite (MPa) is more than any of the individual (ferrite/cementite) phases of which it is composed. Mechanical properties, i.e. strength, ductility, toughness and hardness of pearlite, depend on the inter-lamellar spacing. Finer the inter-layer spacing better the mechanical properties.

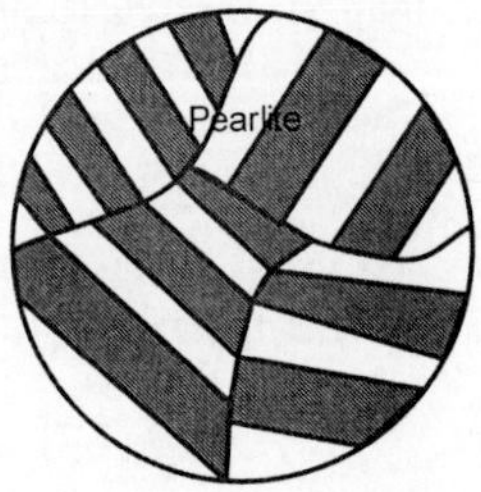

Figure 4.19: Schematic diagram showing optical microscopic features pearlite (mixture of α-Fe & Fe_3C)

Ledeburite

Ledeburite is also a phase mixture of austenite and cementite. This meta-stable phase mixture is produced through a eutectic reaction (4.3% C, 1130 °C). It can be noticed only above the eutectoid temperature on the right side region of Fe-C diagram. The austenite present in ledeburite gets transformed into pearlite below the eutectoid temperature.

4.5.3 Effect of Phases on Mechanical Properties

It is important to note that every phase or phase mixture has its mechanical properties. Few phases are very soft (ferrite, austenite), while a few are extremely hard and brittle (cementite, martensite). The austenite and martensite are met-stable phases and developed steels of specific composition and cooling conditions. Therefore, variations in proportions / relative amounts of these phases affect the mechanical properties of steel. The percentage of pearlite increases (at the cost of ferrite) linearly with an increase in carbon content (Fig. 4.20). Since ferrite has low strength, it is soft and ductile, while pearlite is hard, strong but of poor ductility and toughness, therefore increase in the percentage of pearlite also increases strength, hardness but reduces the ductility and toughness (Fig. 4.21). Cementite appears as an individual constituent only above eutectoid composition (0.8% C). It forms a network along the grain boundary of pearlite depending upon carbon content. Complete isolation of pearlite colonies with the cementite (because of a continuous network of cementite) decreases tensile strength and ductility because mechanical properties of the alloy / steel, largely, depend upon the properties of phase, which is continuous in the alloy. An increase in carbon content above the eutectoid composition reduces the strength and ductility because in hypereutectoid steel, a network of cementite is formed along the grain boundaries of pearlite and cementite has a very low tensile strength (3.0MPa). The network of cementite is discrete for low carbon content (0.8-1.0%) and is continuous for further higher carbon content. A continuous network is more harmful to the mechanical properties than the discrete network of the cementite.

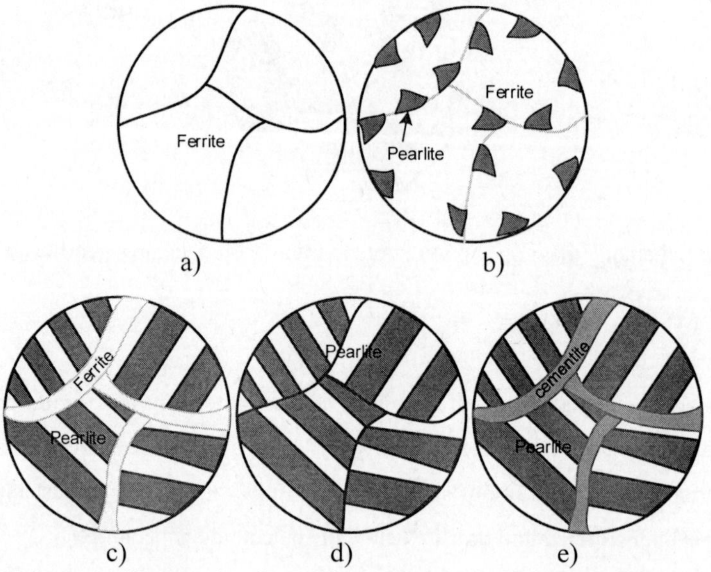

Figure 4.20: Schematic diagram showing optical microscopic features steel with varying carbon content a) 0 % C, b) 0.2 % C, c) 0.6% C, d) 0.8% C and e) 1.2 %C

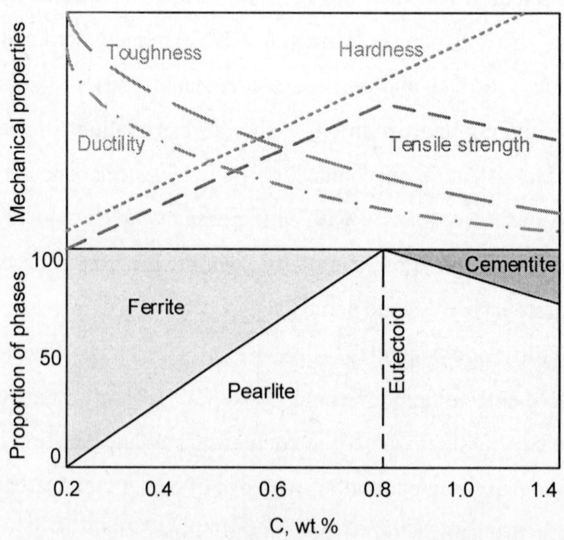

Figure 4.21: Schematic diagram showing variation in micro-constituents in plain carbon steel and mechanical properties as a function of carbon content

Apart from alloying element in steel including carbon and other solid-solution elements, the strength of a ferritic steel is influenced by its grain size as per the Hall-Petch relationship:

$$\sigma_y = \sigma_o + k\,(d^{-1/2})$$

Where σ_y is the yield strength (in MPa), σ_o is a constant, k is a constant, and d is the average grain size (in mm).

The properties of fully pearlitic steels depend on inter-lamellar spacing between the ferrite & cementite lamellae (x), and the size of pearlite colonies. One typical empirical relation between yield strength and inter-lamellar spacing is as under:

$$\sigma_y = -85.9 + 8.3\,(x^{-1/2})$$

Where σ_y is the 0.2% offset yield strength (in MPa) and x is the inter-lamellar spacing (in mm).

4.6 Phase Transformation and its Relevance

Steel with 0.8% carbon (precisely 0.77% C) is known as eutectoid steel. Steels with less than 0.8% carbon are called hypereutectoid steels, and those with more than 0.8% and less than 2% carbon are called hypereutectoid steels. Fe-C systems having carbon of more than 2% are called cast irons. Cast iron having 4.3% carbon is known as eutectic cast iron. Cast irons with less than 4.3 % carbon are known as hypoeutectic cast iron, and those with more than 4.3% are called hypereutectic cast iron. The above classification of steels is based on steel composition with regard to eutectoid reaction point (0.77 wt. % C). The eutectoid reaction point for plain carbon steel occurs at 0.77% C and 727 °C per Fe-C phase diagram. However, the location of the eutectoid point in Fe-C changes with the presence of alloying element (Fig. 4.22 a, b). Eutectoid reaction point moves upward on the right side with the addition of ferrite stablishing element (Cr, V, W etc.) and the same goes down on the left side in the presence of austenite stabilizing element (Ni, Mn, Si etc.). The sifting of the eutectoid point indicates that alloying element can increase/decrease the temperature and carbon content for eutectoid reaction (pearlite to austenite transformation during heating and austenite to pearlite transformation during cooling). Conversely, alloying of steel affects the design of the heat treatment cycle (maximum heating temperature, holding time and cooling rate) to achieve the desired microstructure and mechanical properties. Further, alloying elements forming solid solution in

Fe (steel) affect the yield strength and hardness. However, the relative effect of these alloying elements on yield strength varies significantly (Fig. 4.22c). A common categorization of plain carbon steel (based on carbon content) is low carbon steel (<0.25 wt. %), medium carbon steel (0.25 to 0.5 wt. % C) and high carbon steel (> 0.5 wt. % C). Steel for commercial applications are designed mostly with carbon maximum up to 1.0 wt. %.

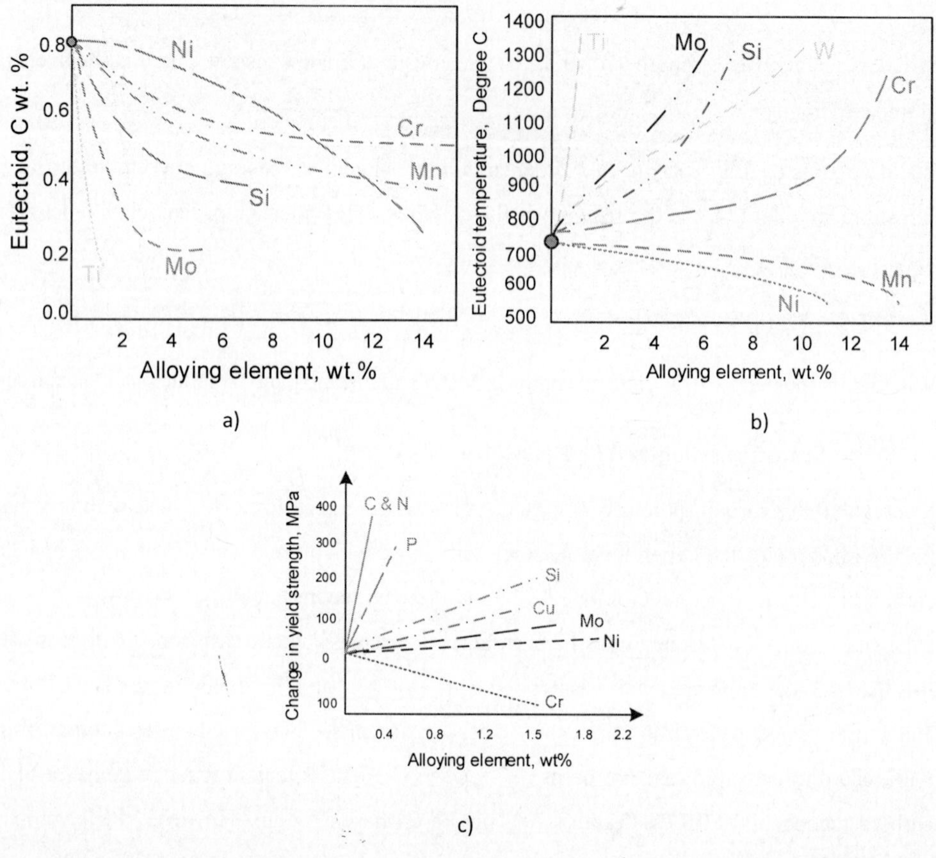

Figure 4.22: Effect of alloying elements on a) eutectoid carbon wt.% and b) eutectoid temperature and c) yield strength

4.7 STEEL

4.7.1 Hypoeutectoid Steel

Transformation of 0.25%C steel during the cooling from a molten state to room temperature is shown by (blue) vertical line "A-H" through the steel composition point (C %). It can be observed that no phase transformation takes place until the temperature of liquid metal goes down from point 'A' to 'B' on the liquidus (Fig. 4.23). At point 'B', solidification starts and first of all, δ-Fe is formed. Further, a decrease in temperature results in the formation of more and more solidification of δ-Fe. The composition of δ-Fe changes along the solidus (green line), whereas the composition of liquid metal changes along the liquidus (blue line). Liquidus shows that carbon content in the molten metal increases with the reduction in temperature until it reaches 0.53 wt%. The liquid metal (0.53% C) and solid δ-Fe (0.1%C) at 1495^{0}C corresponding to the point "C" through peritectic transformation reaction produces austenite (γ-Fe). Thus, δ-Fe is a result of a pro-peritectic reaction and post-peritectic reaction (between liquid and solid δ-Fe), another type of solid, i.e. austenite, is formed.

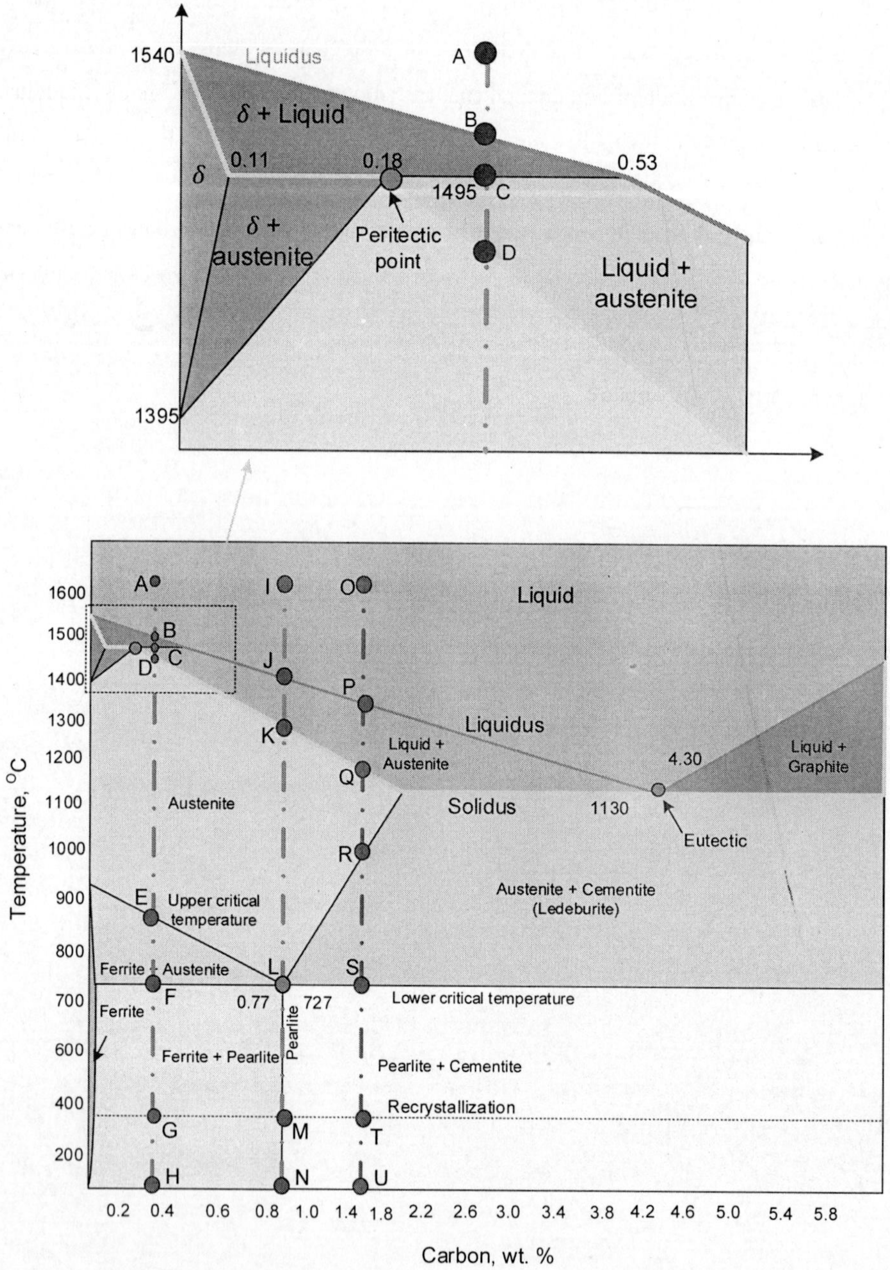

Figure 4.23: Phase transformation occurring in Fe-C diagram for hypoeutectoid steel, eutectoid and hypereutectoid steel on cooling from molten state to room temperature

On completion of solidification at "D", steel attains the complete austenitic state thereafter, no further phase transformation takes place during cooling until the composition line intersects the upper critical temperature line at "E". At about 850^{0}C transformation of austenite into the pro-eutectoid ferrite begins. Further, the transformation of austenite into ferrite continues with reduction in temperature until lower critical temperature (eutectoid temperature) is attained at "F". During the transformation of austenite in inter-critical temperature range, composition of austenite changes continuously along the upper critical temperature line as carbon content in austenite increases. As lower critical temperature (eutectoid temperature) is attained, carbon content in austenite increases to 0.8%, i.e. eutectoid composition. At this temperature, austenite of eutectoid composition transforms into pearlite as discussed above at point "F". Proeutectoid phase, i.e. ferrite, is formed first from austenite in hypoeutectoid steel, and then austenite transforms into pearlite. Hence, ferrite is found along the grain boundaries of pearlite. Continuity of the ferrite network depends upon the carbon content. Reduction in carbon content increases the continuity of the ferrite network. An increase in carbon content reduces the amount of ferrite and increases the amount of pearlite (Fig. 4.24). Further, a reduction in temperature from lower critical temperature / eutectoid temperature to till room temperature "F-H", does not cause any phase transformation.

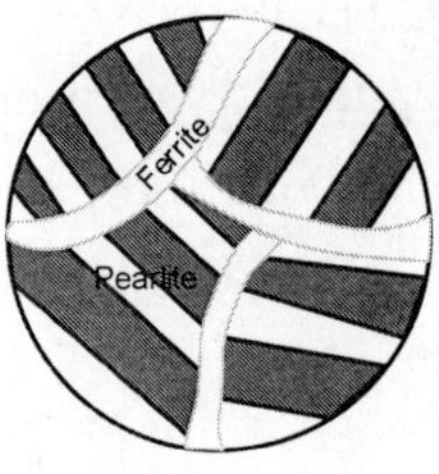

a)

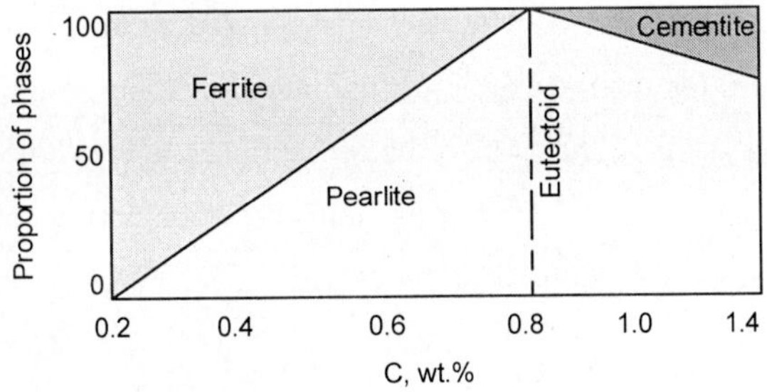

b)

Figure 4.24: Schematic diagram showing a) optical microscopic features of hypoeutectoid steel and b) variation in phase of plain carbon steel with carbon content

4.7.2 Eutectoid steel

Transformation of eutectoid steel (0.77 wt.% C) into various phases on cooling from molten state to the room temperature can be shown by drawing a vertical line (dark yellow) through the composition point (0.77% C) on the X axis (called composition line). No phase transformation takes place until the temperature of liquid metal goes down from point '' to 'J' on the liquidus (Fig. 4.23). At the point 'J', solidification starts and first of all, the austenite crystals are formed. Further temperature decrease results in more and more austenite formation. The composition of austenite being formed change along the solidus "green line" composition of liquid changes follows the liquidus (red line). During this transformation, carbon content in the molten metal increases with the reduction in temperature until it goes down to the point 'K'. On completion of solidification, steel attains the austenitic state thereafter, no phase transformation takes place until it achieves eutectoid temperature (727^{0}C) corresponding to the point "L".

Reduction in temperature decreases the solubility of carbon in austenite. Therefore, austenite becomes saturated with carbon as temperature goes below the eutectoid temperature; hence austenite rejects the excess carbon leading to the nucleation of cementite along the grain boundary of austenite. As a rejection of carbon from austenite continues, cementite layer grows toward the centre of austenite grain. Therefore, a very small zone, which is depleted of carbon, is formed on both sides of the cementite plate. As the concentration of carbon in the "carbon depleted zone" reduces to such an extent when it dissolves in iron with BCC structure, austenite transforms into ferrite. The thickness of the cementite or ferrite plates depends on the diffusion rate and time available for transforming austenite into pearlite. These transformations occur through nucleation and growth mechanisms based on the diffusion of atoms. A low cooling rate increases the time available for the transformation, resulting in a thicker cementite and ferrite layers (Fig. 4.25). This process continues until whole austenite is transformed into pearlite. Thereafter, no further phase transformation during cooling from "L-N" of eutectoid steel occurs.

Figure 4.25: Schematic diagram showing optical microscopic features of eutectoid steel

4.7.3 Hypereutectoid steel

The transformation of hypereutectoid steel is similar to that of hypoeutectoid steel except that hypereutectoid steel results in cementite as proeutectoid phase instead of ferrite. The phase transformation in hypereutectoid steel can be observed from the composition "violet" line in the Fe-C phase diagram. In the case of hypereutectoid steel, no phase transformation takes place until the temperature of liquid metal goes down from point 'O' to 'P' on the liquidus "red line" (Fig. 4.23). At the point 'P', solidification starts and first of all austenite is formed. Further, a decrease in temperature results in the formation of more and more amount of austenite. The composition of austenite being formed during the solidification changes along the solidus "green line" whereas composition of liquid metal goes along the liquidus "red line". It shows that carbon content of molten metal increases with the reduction in temperature until it goes down to the point 'Q'. On completion of solidification, steel attains the fully austenitic state and after that no phase transformation takes place until it attains temperature corresponding to the point 'R'. A temperature reduction decreases carbon's solubility along the upper critical temperature line "solvus line". Reduction in temperature of austenite makes it supersaturated with C; therefore rejection of carbon starts at the grain boundaries of austenite which in turn appears as cementite as pro-eutectoid phase. Further, decrease in temperature results in more cementite formation from the austenite, whose composition changes along the upper critical temperature line. The dissolved carbon content in austenite decreases with a reduction in temperature. As the eutectoid temperature is attained, the carbon content in austenite decreases up to the eutectoid composition (0.8%C). At eutectoid temperature, austenite again becomes supersaturated with carbon; hence austenite rejects the excess carbon, which leads to cementite nucleation along the grain boundary. At this temperature, remain austenite of eutectoid composition transforms into pearlite, as discussed above. The proeutectoid phase, i.e. cementite, is formed first from austenite in hypereutectoid steel; subsequently, austenite transforms pearlite. Hence, cementite is found along the grain boundaries of pearlite (Fig. 4.26).

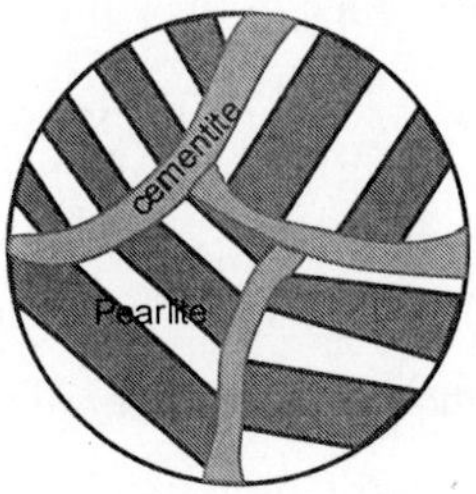

Figure 4.26: Schematic diagram showing a) optical microscopic features of hypereutectoid steel

4.8 Cast Irons

The Fe-C system with a carbon content of more than 2.0 wt.% is called cast iron. Fe-C system with 4.3% carbon forms eutectic; therefore, it is called eutectic cast iron. A cast iron with carbon content from 2.0 to 4.3 wt.% is known as hypoeutectic cast iron, and those with more than 4.3% carbon are called hypereutectic cast iron. Carbon content greater than 4.3 wt.% forms graphite as per the Fe-C phase diagram (Fig. 4.27). The carbon present in cast iron can be in different forms like free form, i.e. graphite or combined form iron carbon, while the matrix can be ferritic, pearlitic, martensitic, bainitic or a combination of these. Martensite is a supersaturated solid of carbon solution in iron with body-centered tetragonal structure while bainite is similar to pearlite but has a very intimate mixture of ferrite and cementite.

Physical metallurgy of cast irons

Cast iron is an alloy of iron-carbon system which melts at the lowest temperature. The low melting temperature of cast iron and other attractive characteristics like low thermal expansion coefficient make it easy to cast. Therefore, it is called cast iron. High carbon content in cast iron leads to high hardness and compressive strength but low toughness and poor ductility. However, the mechanical properties of cast iron significantly depend on two aspects a) form in which carbon is present in cast iron and b) morphology in carbon/its compound. Carbon in cast iron can be present in free or combined form.

Carbon in free form in cast iron appears as graphite which is soft and acts solid lubricant. The free form of carbon (graphite) morphology can be nodules, flakes and irregular shape in nodular, grey and malleable cast iron, respectively. Carbon can also be present in the combined form "iron carbon or alloy carbide". The most common form of cast iron in the combine form of carbon (iron carbide) is white cast iron. The carbon in combined form results in very high hardness and abrasive wear resistance but almost negligible toughness and ductility.

Alloying elements and cooling rate experienced by cast iron during the solidification significantly affect the type of cast iron. The presence of alloying elements like W, V, Cr increasing hardenability at high cooling in variably produces white or alloy cast iron. Few alloying element like Si promote the graphitization to produce grey cast iron. An alloying element like Mg in molten cast iron modifies morphology of graphite flake to nodules leading to the nodular / spheroidal cast iron. Heat treatment of grey cat iron produces the irregular shape of graphite to make malleable cast iron.

4.8.1 Eutectic Cast Irons

Transformation of eutectic cast iron into various phases on cooling from molten state to the room temperature can be noticed by drawing a vertical 'yellow" line on the Fe-C diagram through the composition point (at 4.3%C) as shown in Fig. 4.27. No phase transformation takes place until temperature of liquid metal goes down from point g to h on the liquidus "red line". At point "h", solidification starts and completes at constant temperature (1130^{0}C) through eutectic reaction leading to transformation liquid metal in to ledeburite. Ledeburite is a mixture of austenite and cementite. Austenite formed by eutectic reaction has 2.0 wt.% C. Solubility of carbon in γ-Fe with F. C. C. structure decreases from 2% to 0.8% with the reduction in temperature from 1130^{0}C to 727^{0}C. Hence, austenite rejects excess carbon, leading to the increasing amount of cementite along the grain boundary of the austenite. As the eutectoid temperature is reached, the carbon content in austenite decreases up to 0.8 wt.% at point "i". As discussed above, austenite of eutectoid composition transforms into pearlite at this temperature. Thereafter, no further phase transformation occurs until room temperature "J".

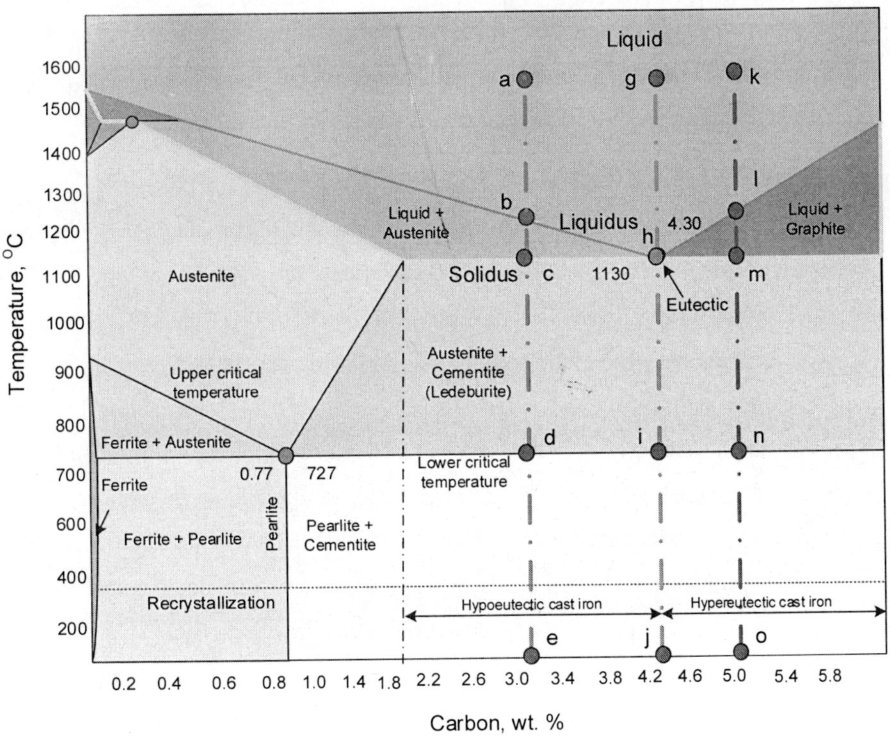

Figure 4.27: Phase transformation observed through Fe-C diagram in hypoeutectic cast iron, eutectic cast iron and hypereutectic cast iron on cooling from molten state to room temperature

4.8.2 Hypoeutectic Cast Iron

Cast irons having a carbon content less than 4.3% are called hypoeutectic cast irons. These cast irons on cooling from the molten state, proeutectic austenite crystallizes first; as solidification progresses, more and more amount of austenite is formed, whose composition changes along the solidus while the composition of liquid metal moves along the liquidus (refer to blue composition line a-b-c). As the temperature goes down to point "c", liquid metal attains the eutectic composition, and subsequently, the whole of the liquid metal gets solidified as ledeburite at constant temperature (1130^{0}C) through eutectic transformation. Austenite present in alloy in any form ultimately transforms into pearlite at eutectoid temperature, as discussed above.

4.8.3 Hypereutectic Cast Iron

Cast irons with more than 4.3% carbon content are called hypereutectic cast irons. Phase transformation of hypereutectic cast irons (refer violet composition line) is similar to that of hypoeutectic cast irons, except that in the case of hypereutectic cast iron pro-eutectic phase is graphite.

UNIT SUMMARY

This unit presents all the relevant topics to understand the phase diagram and related metallurgical transformations including Gibbs Phase rule, lever rule, and various types of phase diagrams of the binary alloys, isothermal transformation reactions like eutectic, peritectic, eutectoid, peritectoid, and monotectic reaction. A systematic understanding of the various phase transformations in the binary alloy can easily be understood using binary phase diagrams. Three types of binary phase diagrams have been included in this unit: binary alloys completely soluble both in solid and liquid state, binary alloys completely soluble in liquid state and partially soluble in solid state, binary alloys completely soluble in liquid state but insoluble in solid state. Additionally, the iron-carbon phase diagram, various isothermal transformation reactions observed in iron-carbon diagram, and phase transformations taking place in iron-carbon system during cooling from molten state to room temperature in hypoeutectoid, eutectoid, hypereutectoid steels besides hypoeutectic, eutectic and hypereutectic cast irons have been presented.

EXERCISE

Questions for self-assessment

1. A type of solid solution formed in the case of Nickel-copper alloy is
    a. Substitutional
    b. Interstitial
    c. Intermetallic
    d. All of these
2. An interstitial solid solution is formed when solute atoms
    a. Replace the atoms of the solvent in the lattice structure
    b. Accommodated in the spaces between the solvent atoms
    c. Occupy the corners atoms of unit cells of solvent metal matrix
    d. Form intermetallic compounds with the solvent
3. A binary alloy that shows complete solubility in the liquid state and complete insolubility in the solid state
    a. Fe-C
    b. Al-Si
    c. Bi-Sn
    d. Ti-Al
4. The phase diagram of a binary alloy helps to determine
    a. The temperature at which the phase transformation will be occurring
    b. The phases present at the room temperature
    c. The temperature at which solidification will be completing
    d. All of these
5. An isomorphous binary alloy system shows
    a. Complete solubility alloying elements in the solid and liquid state
    b. Complete solubility of the alloying elements in the liquid state and partial solubility in the solid state
    c. Complete solubility of the alloying elements in the liquid state and insoluble in the solid state
    d. All of these

6. Peritectic reaction in iron carbon diagram results in
   a. Ferrite
   b. Austenite
   c. Pearlite
   d. Cementite
7. The eutectic in the iron-carbon system is formed at
   a. 2% carbon
   b. 4% carbon
   c. 4.3 % carbon
   d. 6.67% carbon
8. Which one of the following isothermal transformations occurs in the solid state
   a. Peritectic reaction
   b. Eutectic reaction
   c. Monotectic reaction
   d. Eutectoid reaction
9. An iron-carbon system with 2.2% carbon can be classified as
   a. Hypoeutectoid Steel
   b. Hypereutectoid Steel
   c. Hypoeutectic cast iron
   d. Hypereutectic cast iron
10. White cast iron has carbon in the form of
    a. Nodules of the graphite
    b. Flakes of graphite
    c. Regular shape of graphite
    d. Iron carbide
11. The maximum solubility of carbon in Fe at room temperature is found in
    a. α-Ferrite
    b. Martensite
    c. Austenite
    d. Delta-ferrite
12. In Gibbs phase rule number of components shows
    a. Number element present in a system
    b. Number of phases
    c. Number of variables

d. All of these

Answers of Multiple Choice Questions

Key for MCQ: 1 a, 2 b, 3 c, 4 d, 5 a, 6 b, 7 c, 8 d, 9 c, 10 d, 11c, 12a

Short and Long Answer Type Questions

1. How does the number of components in an alloy system affect the number of phases, explain using Gibbs Phase Rule.
2. Explain the Gibbs Phase rule.
3. Differentiate interstitial and substitution of solid solutions.
4. Enlist information that can be obtained from a phase diagram.
5. Explain lever rule using suitable example.
6. What is the eutectic point?
7. Explain the mechanism of the eutectic reaction and the morphology of its products.
8. Explain the following isothermal transformation reactions
    a. Monotectic reaction
    b. Peritectic reaction
    c. Eutectic reaction
    d. Eutectoid reaction
    e. Peritectoid reaction
9. Explain the phase transformations in the isomorphous Ni-50%Cu alloy system with the help of a relevant phase diagram
10. Draw and label the iron-carbon diagram and show various isothermal transformation reactions.
11. With reference to the Fe-C diagram, explain the following isothermal transformation reactions
    a. Peritectic reaction
    b. Eutectic reaction
    c. Eutectoid reaction
12. Describe the phase transformations which will be occurring on heating of 0.3% carbon steel from room temperature to 1100 degree centigrade.
13. On what basis the different types of cast irons are classified?
14. Enlist the factors affecting the microstructure of cast iron.

Numerical Problems

1. Using the Lever rule determines the proportions of various phases for the following iron-carbon system
    a. Ferrite & cementite type in 0.4% Carbon Steel
    b. Pearlite & cementite type in 0.8% Carbon Steel
    c. Austenite and Ledeburite after eutectic reaction in 3.0% cast iron at 1130 degree centigrade
2. Fe-C phase diagram showing eutectoid and eutectic at 0.8% and 4.3% C. Using lever rule determine
    a. Fraction of ferrite and pearlite in 0.3 % C steel at room temperature
    b. Fraction of pro-eutectic austenite and liquid metal in Fe-2.3%C cast iron at 1130 °C
    c. Fraction of pro-peritectic Fe and liquid metal in Fe-2.3%C cast iron at 1130 °C
3. Using image analysis of the following scanning electron microscopic image (showing) microstructure of carbon steel quantify proportion of pearlite (white) and ferrite (dark) and using lever rule calculate C wt.% in steel.

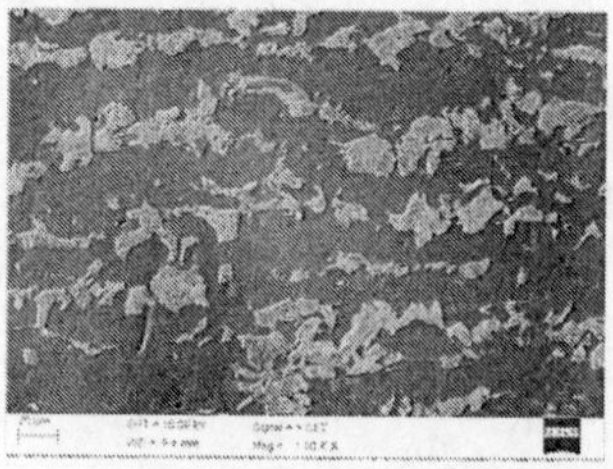

PRACTICAL

1. Polish, etch and study microstructure of carbon steel or mild steel using optical microscopy.
2. Develop a cooling curve (plot temperature vs time diagram) during the solidification of metal and alloy using a suitable thermal couple, and write your observations
3. Using image analysis (like Image J open source software), determine the grain size and % of pearlite and ferrite from the microstructure of carbon steel observed at 200 / 400 magnification

KNOW MORE

Explore the microstructure and Mechanical properties of amorphous metals. The methods used to manufacture the components of amorphous metals. Learn underlying principles in the development of amorphous metals. Rapid solidification technique to manufacture the amorphous foils of aluminium. Compare the structure and properties of the crystalline and amorphous metals. Using the cooling curves, develop the phase diagrams for suitable binary alloy system.

REFERENCES AND SUGGESTED READINGS

Suggested resources for further reading/learning

1. Callister, Materials science and engineering, Wiley, Publication (2014)
2. S D Avner, Introduction to Physical Metallurgy, McGraw Hill, Publication (2002)
3. GE Dieter, Mechanical metallurgy, McGraw Hill, Publication (1989)
4. D K Dwivedi, Fundamentals of Metal Joining, Springer nature, Singapore (2022)
5. D K Dwivedi, NPTEL Course, Fundamentals of Manufacturing Processes, https://archive.nptel.ac.in/courses/112/107/112107219/
6. Ratna Kumar, NPTEL Course "Basics of Materials Engineering" https://onlinecourses.nptel.ac.in/noc20_me78/preview
7. Structure/Property Relationships in Irons and Steels, ASM International, Second Edition J.R. Davis, Editor, (1998), 153-173

Dynamic QR Code for the Unit 4

Unit 5

Heat treatment of steel

Microstructure, TTT and CCT Diagram, Annealing, Normalizing, Hardening, Case Hardening

Unit Specific / Learning Objective

The objective of this unit is to talk about the following aspects

- To introduce the isothermal transformation diagram and continuous cooling transformation diagram for the iron-carbon system
- To learn the application of Fe-C diagram, TTT diagram, and CCT diagram for designing the heat treatment of steel
- To develop an understanding of the selection of suitable heat treatment for achieving the desired mechanical properties in a steel
- To introduce the concept of case hardening and common methods used for case hardening of steels
- To learn about the heat treatment cycles to be used for common heat treatment of steel
- To understand the way microstructural modification of steel by heat treatment helps to develop the desired mechanical properties
- To introduce typical heat treatment like in-process annealing to facilitate the manufacturing and stress relieving to improve the mechanical performance

Additionally, a few fundamental questions for self-assessment based on fundamentals are included in this chapter in the form of recall, application, comprehension, analysis and synthesis. Further suggested readings and references are for deep learners and the reader's assistance.

Rationale

Applying controlled heating, soaking, and cooling cycles on metals like steel, and few aluminium, and titanium alloy helps to develop the desired microstructure (in terms of the phase and grain structure). The microstructural modification (using heat treatment) allows change in phases from soft to hard and vice versa and grain size from coarse to fine and vice versa. Such microstructural alterations can be easily achieved in carbon and alloy steels by applying a controlled heat treatment cycle. Therefore, steel heat treatment can be used to harden or soften it as needed. Also, heat treatment relieves residual stresses, reducing hard steel distortion and cracking tendency and improving mechanical properties. Therefore, it is important to learn how heat treatment can help achieve suitable microstructural modification and enhanced mechanical properties.

Pre-Requisites

A course on chemistry: Class XII

Learning outcomes

U5-O1: Ability to select the suitable heat treatment to achieve the desired Mechanical properties of office Steel

U5-O2: Ability to design the heat treatment cycle to develop the desired Mechanical properties in steels varying compositions

U5-O3: Ability to use the TTT, CCT, and Fe-C diagrams for designing the heat treatment

U5-O4: Ability to estimate the microstructural changes under a given set of heating and cooling conditions

U5-O5: Ability to design and apply the suitable method for case hardening of steel to improve the hardness and wear resistance

Unit-4 Outcomes	EXPECTED MAPPING WITH COURSE OUTCOMES (1- Weak Correlation; 2- Medium correlation; 3- Strong Correlation)					
	CO-1	CO-2	CO-3	CO-4	CO-5	CO-6
U5-O1	1	3	2			
U5-O2	1	3	2			
U5-O3	1	3	2			
U5-O4	1	3	2			
U5-O5	1	3	2			

Course Objective

4. Students will be able to identify crystal structures for various materials, and one could understand the defects in such structures.
5. Understand how to tailor the material properties of ferrous and non-ferrous alloys?
6. How to quantify mechanical integrity and failure in materials?

5.1 Introduction

One of the main objectives of this subject is to achieve the desired set of mechanical properties by controlling the microstructure of the materials. We have learned about the crystal structure, mechanical properties, phase transformations observed in the metals during the heating and cooling, and the stress conditions leading to the failure of the materials under uniaxial and multiaxial loading conditions. This unit applies those fundamentals to achieve the desired mechanical properties by controlling the microstructure of steel through the controlled cycle of heating and cooling called heat treatment. The heat treatment helps to modify the microstructure and mechanical properties uniformly across the section, changing the bulk material property.

The fundamentals related to steel heat treatment, including the Fe-C phase diagram, isothermal time transformation, and continuous cooling transformation diagrams, have been presented. A good understanding of these fundamentals will help learners design suitable heat treatment cycles per the desired combination of mechanical properties for a given application. The common heat treatment used for achieving the desired Mechanical properties of steel, namely annealing, normalising, quenching/ hardening, tempering, austempering, and martempering, have been explained using suitable schematic diagrams and heat treatment cycles.

Few engineering applications need different properties at the surface from the core. In such cases, the case hardening treatment is applied. The case hardening increases surface hardness without altering the toughness of the core of the material. Thus, case hardening provides hard surfaces with a tough core. Hard surfaces improve wear resistance, while the tough core provides the desired mechanical load-carrying capacity and resistance to impact loading. Further, the typical case hardening treatments such as flame and induction hardening, carburising, cyaniding, nitriding, carbo-nitriding, plasma nitriding, and vacuum-based techniques like ion implantation are presented in the following sections.

5.2 Need of heat treatment

The few metals allow the change in crystal structure and microstructure when subjected to a controlled heating and cooling cycle. This control over the microstructure through heat treatment helps to change the mechanical properties per the application's requirement. During the deformation-based manufacturing of steel, it is required to induce the desired softness and ductility so that it can be processed further without increasing the cracking tendency and brittleness. The process of annealing can easily achieve this. Heat treatment like normalising increases strength and hardness without significantly compromising the ductility and toughness by refining the microstructure. Residual stresses are reduced during manufacturing by forming, casting, and welding. Residual stresses lower mechanical properties like tensile and fatigue strength and increase distortion tendency. Therefore, heat treatments such as stress relieving and annealing are performed to reduce the residual stress. Sometimes, it is required to increase the strength and hardness of steel then hardening treatment is done. The hardening heat treatment increases the hardness and strength but at the cost of toughness and ductility, making the steel brittle. Therefore, hardened steels are tempered to induce some toughness and relieve the residual stresses induced during the hardening.

5.3 Physical Metallurgy of Heat Treatment

5.3.1 Principle

Steel and cast iron offer an attractive feature of changing the microstructure through the control cycle of heating and cooling. On heating above a certain critical temperature (as per the composition), phases (ferrite, pearlite, cementite, bainite, and martensite) present in the Fe-C system at room temperature transforms into the metastable phase austenite. The formation of homogeneous austenite requires heating at a high temperature for a certain time, called soaking time. The austenite being a meta-stable transforms (except in austenitic steel) into various phases (from relatively soft ferrite, pearlite, bainite to very hard and brittle phases an compound like cementite, martensite and other complex phases) depending upon the applied controlled cooling rate and composition of the Fe-C system. Therefore, the application of appropriate cooling results transformation of austenite into different phases, which forms a basis for various heat treatments like annealing, normalising, hardening, austempering, martempering etc. It is relatively easier to achieve the homogeneous austenitic state in carbon, and micro-alloyed steel, followed by controlled cooling to achieve the desired microstructure and mechanical properties. Further, the few heat treatments rely on transforming hard phases like martensite and cementite into soft phases to realise the desired combination of strength, toughness, and ductility (through tempering) and improving machinability (through spheroidization).

5.4 Heat treatment cycle

The heat treatment of metals involves heating to a high temperature, holding at a high temperature for a certain time for homogenisation followed by control cooling to achieve the desired microstructure and a combination of mechanical properties (Fig. 5.1). The heat treatment cycle shows the variation in temperature as a function of time during the heating, holding, and cooling stages of heat treatment. Relatively low temperatures and long-time processes characterise the typical heat treatment cycle compared to other manufacturing techniques like casting and welding.

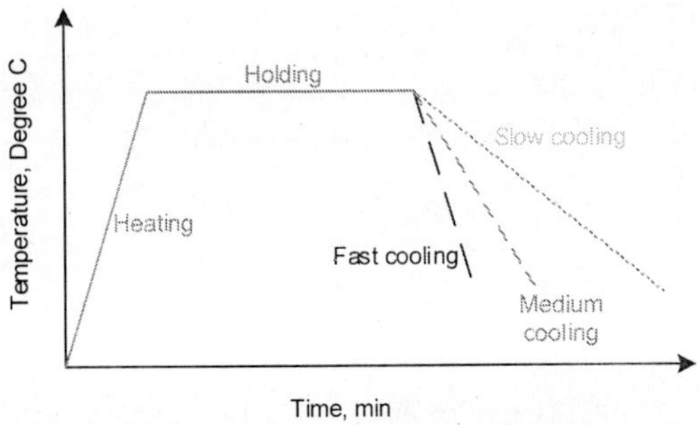

Figure 5.1: Schematic diagram of heat treatment thermal cycle

5.5 Austenitization

Transformation of soft phases (like ferrite, pearlite, and cementite) present at room temperature in a Fe-C system into homogeneous austenite is needed to perform heat treatment like annealing, normalising, quenching, austempering, and martempering. The heating of plain carbon steel above the upper critical temperature (as per the iron-carbon phase diagram) results in austenite (Fig. 5.2). Austenite must be of uniform composition to achieve a similar phase transformation during the cooling regime of heat treatment. Austenite of heterogeneous composition after austenitization results in different kinds of phases during the cooling, which may result in undesired microstructure and mechanical properties. Austenite of homogeneous composition can relatively be easily obtained during the heating of plain carbon steel than alloy steel. The presence of the alloying element in alloy steel forms hard and stable phases and compounds that need higher temperatures and longer soaking time for thermal decomposition to produce homogeneous austenite. Further, the location of upper and lower critical temperatures is significantly influenced by the alloying elements, as evident from the effect of alloying elements on eutectoid temperature.

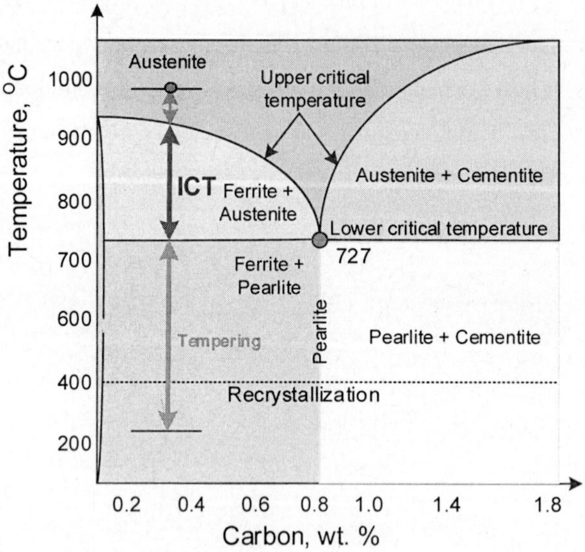

Figure 5.2: Selected region of Fe-C diagram to show a selection of heating temperature for heat treatment of plain carbon steel

5.6 Heating temperature & soaking time role of Fe-C diagram

The heating rate for heat treatment of complex geometry components must be optimum. A very high heating rate due to the differential temperatures in different zones increases the distortion tendency. On the other hand, a low heating rate increases the time required to complete the heat treatment cycle and reduces productivity. The maximum heating temperature is based on the type of heat treatment to be performed. For example, the maximum heating temperature for tempering hardened steel is below the lower critical temperature, while the maximum heating temperature for the heat treatments like annealing, normalising, hardening, austempering, and martempering is usually above the upper critical temperature. The iron-carbon diagram is useful for determining the lower and upper critical temperatures and identifying the maximum heating temperature during the heat treatment of plain carbon steels. In general, an increase in carbon content in plain carbon steel and the presence of austenite stabilising elements in micro-alloyed steel reduce the critical temperatures to achieve the hundred percent austenitic state. Further, the holding time at the maximum temperature depends on the composition of the steel and section thickness. In general, an increase in section thickness and alloying elements in steel increases the holding time to achieve the homogeneous austenitic state (Fig. 5.3). While increasing of heating temperature reduces soaking time to achieve a homogeneous austenitic state.

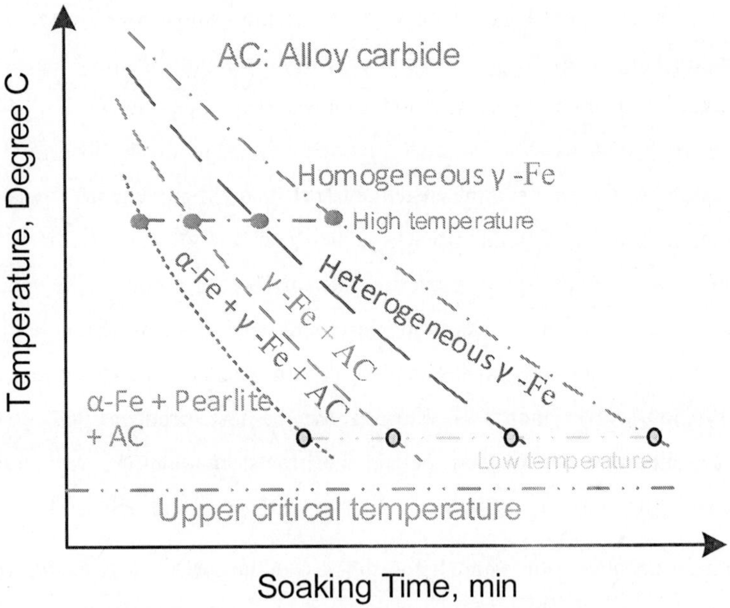

Figure 5.3: Schematic diagram showing how soaking time affected obtaining a homogeneous austenitic state with heating temperature and alloy constituents (as per composition) for heat treatment of a steel

5.7 Isothermal transformation (Temperature-Time-Transformation) diagram

The temperature-time transformation diagram shows the transformation of metastable austenite phase at a constant temperature into various phases as a function of time (Fig. 5.4). Therefore, it is also known as an isothermal transformation diagram.

The transformation of austenite into various phases, such as pearlite, bainite and martensite, depends on the transformation temperature. The time to start the transformation of austenite into pearlite or bainite is called the incubation period. The incubation period is minimum at intermediate temperature (near nose at about 550°C), and it is found to be higher both at low (>Ms) and high temperature (<critical temperature) transformation temperatures. This variation in the incubation period as a function of the transformation temperature leads to the C shape of the isothermal transformation diagram.

TTT Diagram for eutectoid C- Steel comprises four lines, two each corresponding to start and finish transformation at different temperatures. This curve has a nose at about 550°C. Austenite transforms into pearlite at a temperature above the nose and below the critical lower temperature. Increasing the transformation temperature above the nose improves the time required to start and end the austenite transformation into pearlite. Similarly, austenite transforms into bainite at a temperature below the nose and above the martensite start temperature (M_s). A decrease in transformation temperature low the nose also increases the time required to start and end the austenite transformation into bainite. Transformation of austenite into pearlite or bainite occurs by nucleation and growth mechanism. The rate of nucleation and growth significantly depends on the transformation temperature. High transformation temperature leads to increased growth and low nucleation rates with long transformation times. Long transformation, in turn, results in a coarse pearlitic structure. Reduction in transformation temperature (above the nose) produces fine pearlite due to a high nucleation rate, low growth rate and short transformation time. Low transformation temperature, therefore, offers high strength and hardness due to fine pearlitic structure.

Austenite at a contact temperature below the nose of the curve (above the Ms Temperature) transforms into bainite. Bainite is a very fine intimate mixture of ferrite and cementite, like pearlite. However, pearlite is a mixture of lamellar ferrite and cementite. Therefore, bainite offers higher strength, hardness and toughness than pearlite. The degree of fineness of bainite also increases with the reduction in temperature, like pearlitic transformation. Bainite formed at high temperatures is called feathery bainite, while that formed at low temperatures near the Ms temperature is called acicular bainite.

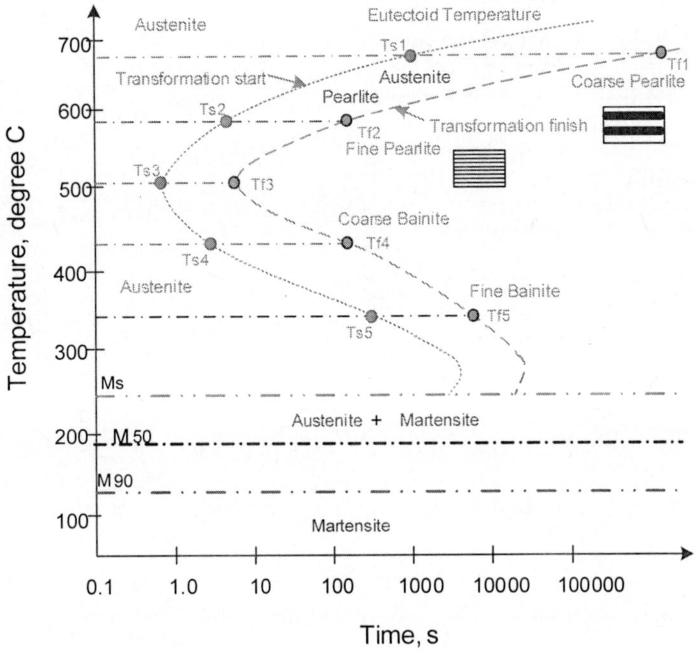

Figure 5.4: Temperature-time transformation or isothermal transformation diagram of eutectoid steel showing phase transformation of austenite (into various phases) at different constant temperatures as a function of time

Austenite at a constant temperature below the martensite start temperature (Ms) transforms into hard and brittle phase called martensite. Martensite is a supersaturated solid solution of carbon in iron having body centred tetragonal (B.C.T) structure. This is a-thermal transformation and occurs through the diffusion-less process. Rapid quenching/cooling from upper critical temperature to a temperature below the 'Ms' prevents any atomic diffusion. Carbon atoms presented in steel are accommodated within the FCC unit cell of austenite at high temperatures and should get rejected at low temperatures on rapid cooling due to the reduction in solid solubility of carbon in iron. But direct cooling a very low temperature (below 'Ms') diffusion is prevented, leading to the formation of a supersaturated solid solution of carbon in iron having BCC structure. This super-saturation of carbon in iron distorts BCC lattice structure and makes it BCT. Due to distortion in lattice from BCC to BCT, the c/a ratio becomes more than 1. Degree of distortion is measured in terms of c/a ratio. This ratio depends upon the carbon content. An increase in carbon content up to 0.8%, increases the c/a ratio. The c/a ratio can be directly related to the increase in hardness, as there is a linear relationship between the two (carbon content and hardness) up to 0.8% carbon content (Fig. 5.5).

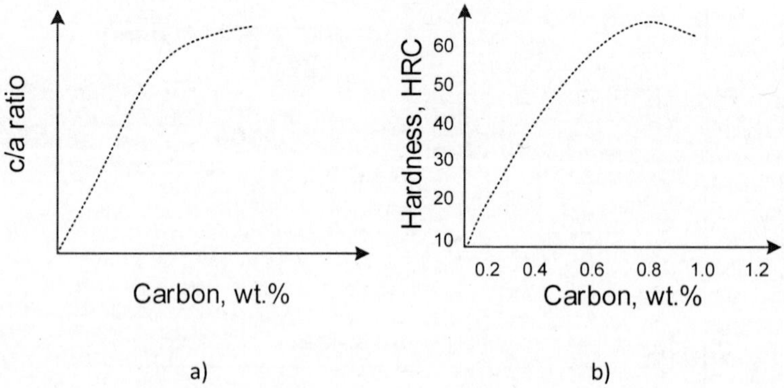

Figure 5.5: Schematic diagram effect of carbon content on a) c/a ratio and b) hardness of steel

Location of austenite to pearlite, bainite and martensite "start transformation lines" and nose depends on alloy composition, homogeneity of austenite and grain size. For each composition, there is one unique TTT diagram. Steel other than eutectoid composition will have one more line initiating from the nose in the TTT diagram corresponding to the austenite transformation into proeutectoid phase. In the case of hypoeutectoid steel, first austenite forms ferrite as a proeutectoid phase, subsequently remaining austenite transforms into pearlite, while the proeutectoid phase in the case of hypereutectoid steel is cementite (Fig. 5.6). Similarly, martensite transformation start and finish (Ms and Mf) temperature lines are affected by alloy composition and homogeneity of austenite. The addition of alloying elements lowers these temperatures.

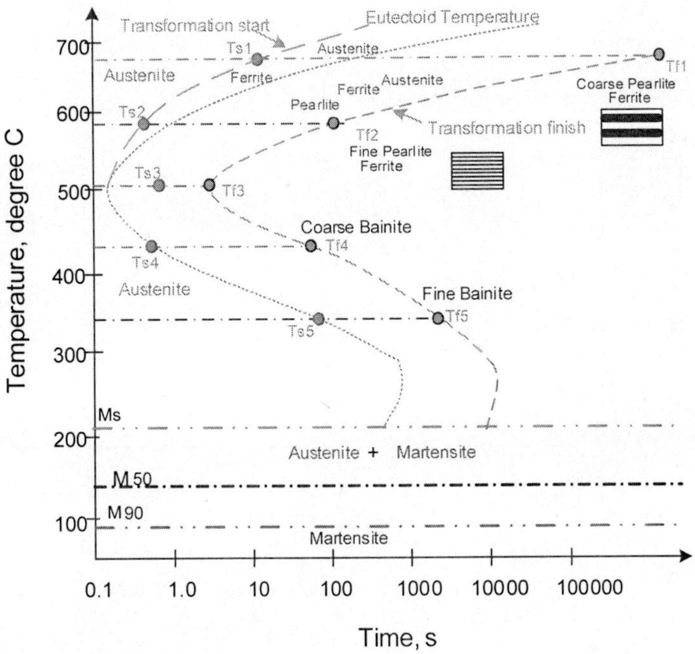

Figure 5.6: Temperature-time transformation or isothermal transformation diagram of hypo-eutectoid steel showing phase transformation of austenite (into various phases) at different constant temperatures as a function of time

Variations in carbon and other alloying elements in steel shifts the nose of TTT diagram. Reduction in carbon (with respect to eutectoid steel) resulting in hypoeutectoid steels, the nose of the curve shifts towards the left while the addition of carbon and alloying element (except a few like cobalt) on the other hand shifts the noses towards the right as compared to that for eutectoid Steel (Fig. 5.7). Similarly, incomplete austenitization reduces alloying elements dissolved in austenite, therefore, effect of alloying element on shifting of the nose of TTT curve is minimised which in turn affects the response of heat treatment to microstructure and mechanical properties of the steel. The location of the transformation line and nose shifting is of great importance in determining the cooling needed to be applied to achieve the desired microstructure for heat treatment because of their effect on critical cooling rate and ease of hardening.

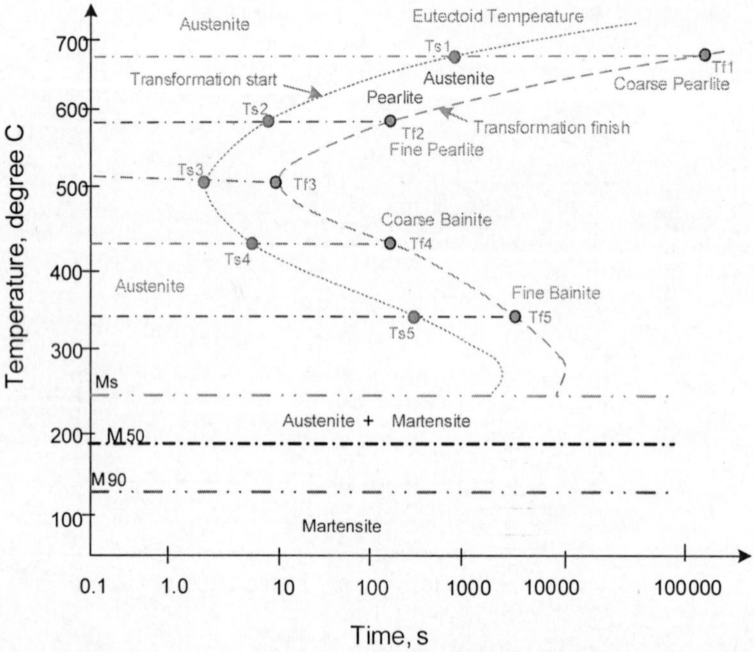

Figure 5.7: Shifting of temperature-time transformation / isothermal transformation diagram to the right due to the addition of alloying element (including carbon) as compared to eutectoid steel

5.8 Controlled cooling and microstructure

Transformation of the metastable austenite on cooling into other phases to some extent can be approximated from the isothermal transformation diagram. To understand these transformations, cooling rate lines are superimposed on the isothermal transformation diagram (Fig. 5.8). However, the location of the transformation start and end lines are found to be different under the actual cooling conditions because the isothermal transformation diagram shows the transformation of austenite into the other phases at a constant temperature only.

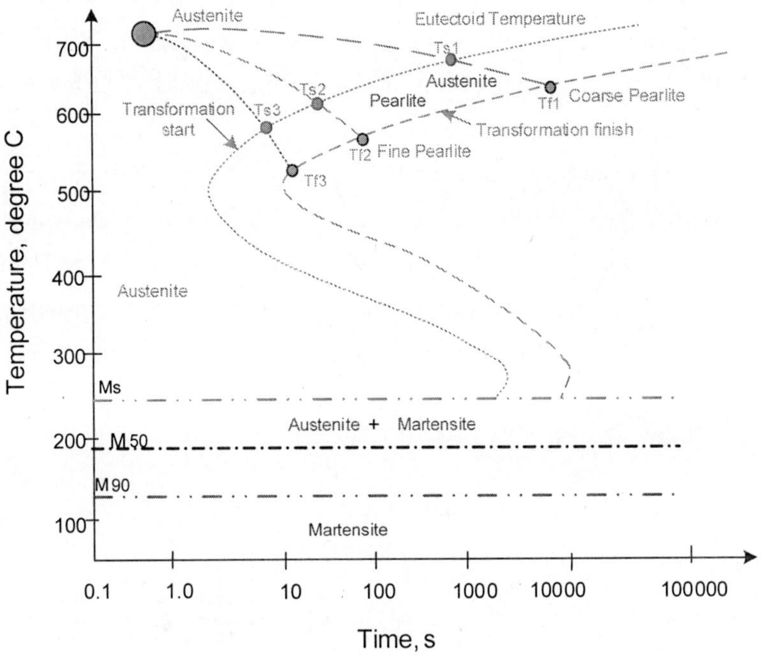

Figure 5.8: Superimposing continuous cooling curve on isothermal transformation diagram to have some idea/approximation of the phase transformation in eutectoid steel under continuous cooling conditions

Therefore, considering the importance of the effect of the cooling rate on the transformation of austenite into the other phases, Steel specific continuous cooling transformation diagrams have been developed. Similar to the isothermal transformation diagram, the continuous cooling transformation diagram also shows the transformation of austenite into various phases as a function of time but under continuous cooling conditions (Fig. 5.9). The continuous cooling transformation diagram, however, is a little different from the isothermal transformation diagram because there is no bainite transformation "region" in the continuous cooling transformation diagram. Additionally, the location of the transformation start and end line in the continuous cooling transformation diagram also slightly differs (the nose of the curve is shifted to the right in the downward direction) as compared to the isothermal transformation diagram.

5.8.1 Continuous cooling transformation (CCT) diagram

The continuous cooling transformation diagram for eutectoid steel has only two lines above Ms, which correspond to the start and end of pearlitc transformation (Fig. 5.9). Effect of cooling rates (CR1 to CR5) on austenite to various phases transformations with the reduction in temperature as a function of time can be seen from Figure. At a very low cooling rate 'CR1', the transformation of austenite into pearlite starts at a high temperature (Ts1) and ends at a little lower temperature (Tf2). Therefore, the transformation of austenite into pearlite occurs over a range of temperatures from Ts1 to Tf2. Since the transformation temperature directly affects the phase and grain structure due to varying nucleation and growth rate. Therefore, the transformation of austenite in the beginning at high-temperature forms a slightly coarser pearlite than at the end stage of low-temperature transformation. Such low cooling rates are used for the annealing of steels to enhance the softness and ductility.

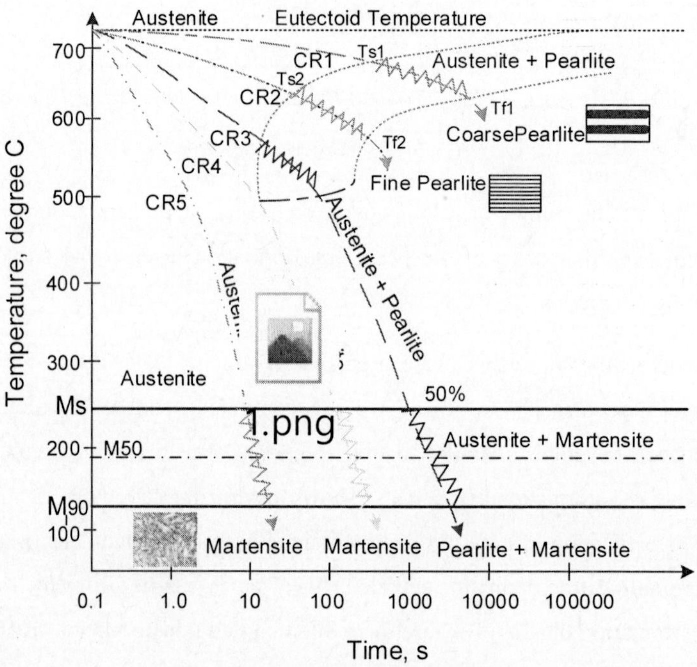

Figure 5.9: Continuous cooling transformation (CCT) diagram of eutectoid steel showing phase transformation of austenite (into various phases) under different cooling rates as a function of time

At a somewhat higher cooling rate (CR2), the austenite transformation into pearlite starts at Ts2 temperature and ends at Tf4 temperature. Therefore, transformation occurs over a range of temperatures from Ts2 to Tf2. Grain size depends on the transformation temperature. High transformation temperature produces coarse grain. Since under continuous cooling conditions, the transformation occurs over a range of temperatures, say Ts2 to Tf2, the grain size also varies accordingly. Therefore, at the start of the transformation, coarser pearlite grains are formed than that at the end of the transformation.

High cooling reduces the effective average transformation temperature, producing a fine-grain structure. Such cooling rates are used in normalising steel. Normalising increases the strength, hardness and toughness due to grain refinement.

Cooling curve 'CR3' is tangential to the 50% transformation line where 50% austenite transforms into pearlite and the remaining 50% austenite is still left to transform. At this stage, the transformation of austenite stops and no further transformation occurs until the temperature goes down below the martensite start temperature (Ms). As austenite crosses the martensite start temperature (Ms) remaining 50% of austenite transforms into the martensite.

Further higher cooling rate 'CR4' curve tangential to the nose of the CCT diagram does not cause any transformation of austenite into pearlite and austenite remains stable until Ms temperature is crossed. Further reduction in temperature transforms the austenite into martensite. Moreover, a complete transformation of austenite into martensite depends on the quenching temperature. If the quenching temperature is below the martensite finish temperature 'Mf', then only the whole austenite will transform into the martensite; otherwise some un-transformed austenite is left in steel as "retained austenite". Retained austenite is comparatively soft; therefore, its presence reduces the hardness of steel. The amount of retained austenite depends on the quenching temperature between Ms and Mf. Lower the quenching / transformation temperature (between Ms and Mf), smaller is the amount of retained austenite. There is non-linear relationship between the amount of austenite transforming into martensite and quenching temperature between Ms and Mf. The minimum cooling rate that is tangential to the nose of CCT curve and ensures the transformation of austenite into martensite (without forming a soft phase like pearlite) is called the critical cooling rate (CCR).

The critical cooling rate depends on the position of the nose. The nose position, as described earlier, depends on steel composition, grain size and homogeneity of austenite. The effect of all alloying elements in steel on the critical cooling rate is taken into account using a single parameter carbon equivalent. The carbon equivalent shows the varying effect of various alloying elements similar to that of carbon and is obtained using the following equation.

High carbon equivalent lowers the critical cooling rate, increasing the ease of hardening. Reducing carbon content increases the critical cooling rate and makes hardening steel difficult. Fine-grained austenite starts transformation into pearlite and other soft phases earlier; therefore, nose of CCT is shifted towards the left. This increases the critical cooling rate. On the other hand coarse-grain, austenite shifts nose toward right, reducing the critical cooling rate and increasing the hardenability. Similarly, inhomogeneous austenite (due to incomplete austenitization) also reduces the transformation time and shifts the nose of the CCT curve to the left, which increases the critical cooling rate.

5.9 Microstructure and mechanical properties of steel

Enormous flexibility associated with the steel to alter the microstructure using heat treatment helps achieve a wide range of mechanical properties for a given steel composition. According to the Iron carbon diagram, steel (up 0.1 to 2.0% C) mainly contains ferrite, pearlite, and cementite at room temperature. Each of these phases has its own set of mechanical properties. Further, the application of different cooling rates also allows variation in grain size and phase structure (fine/coarse ferrite, pearlite, bainite, and martensite). Therefore, the thermal cycle experienced by steel shows significant variation in mechanical properties. Table 5.1 shows different phases targeted to achieve the desired various combination of mechanical properties.

Table 5.1 Mechanical properties and desired phase in steel

S. No.	Mechanical Properties	Phase(s)
1	Softens	High ferrite, low pearlite
2	Ductility	Fine ferrite, pearlite, bainite
3	Yield strength	Fine pearlite, bainite, martensite
4	Ultimate strength	Fine martensite, bainite, pearlite
5	Hardness	Martensite, cementite

6	Toughness	Fine ferrite, pearlite, bainite

5.10 Fundamentals of heat treatment of steel

Steels have the advantage of changing the mechanical properties over a wide range by controlled heating and cooling, i.e. heat treatment (Fig. 5.10). There are three basic steps of heat treatment of steel

- Heating (austenitization)
- Soaking (homogenization)
- Controlled cooling (phase transformation)

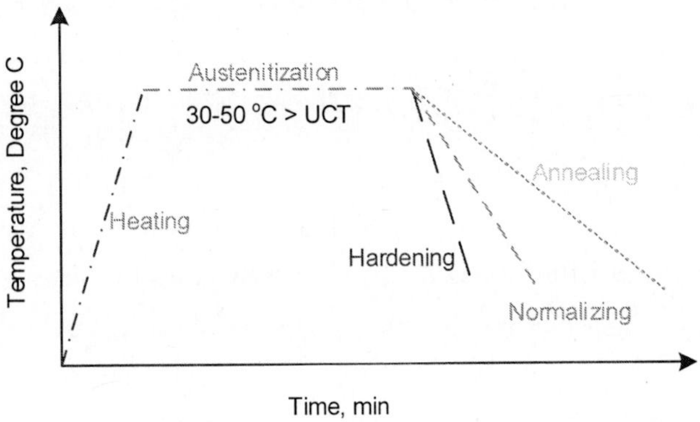

Figure 5.10: Schematic diagram of heating, soaking and cooling cycles for different heat treatments

5.10.1 Heating (austenitization)

In most common heat treatments (annealing, normalizing, hardening, austempering, martempering etc.), steel is heated to transform various phases (pearlite, ferrite, bainite, carbide and alloying element, which are either in or out of the solution) into homogeneous austenite. Austenite formation is necessary because it is a meta-stable phase that can be transformed into different phases per requirements using appropriate controlled cooling. Steel should be heated to an optimum temperature for this transformation. The heating temperature depends upon the alloy composition and the purpose of heat treatment. The heating temperature should be higher than the upper critical temperature (UCT) to get the complete austenitic state steel. UCT is determined by steel composition. For eutectoid steel upper critical temperature is 727^0C, while for pure iron, it is 910^0C. It can be seen from the Fe-C diagram that the upper critical temperature reduces with an increase in carbon content up to the eutectoid composition

(0.8%C) after that, it increases rapidly. Depending upon the composition of the steel, the heating temperature should be about 30^0C-50^0C higher than the upper critical temperature to ensure a complete austenitic state. Heating at too high a temperature (>UCT) causes coarsening of austenite and decarburization and even oxidation of steel, which can adversely affect the mechanical properties of steel.

On heating, the transformation of various phases occurs in reverse order than that of cooling. An increase in steel temperature does not cause any transformation until the temperature goes beyond the lower critical temperature (LCT, i.e. 727^0C). As soon as LCT is crossed, pearlite transforms into austenite at a constant temperature. Thereafter, proeutectoid phase (ferrite or cementite) transforms into austenite with further rise in temperature in the inter-critical temperature region (between LCT & UCT). A complete proeutectoid phase transformation occurs only when steel is heated above the upper critical temperature.

5.10.2 Soaking (homogenisation)

Once the desired heating temperature is achieved, steel components are kept at a high temperature for time (soaking) to realise the homogeneous austenitic state. Soaking time depends upon the alloy composition, heating temperature and dimensions of the components. The presence of alloying elements (Cr, W, V etc.), which form stable inter-metallic compounds (carbides), increases soaking time. High heating temperature reduces soaking time because decomposition and dissolution are promoted at high temperatures due to high diffusivity. The Thicker the section size of the steel component, the longer the soaking time.

5.10.3 Controlled cooling (phase transformation)

The third phase of the heat treatment cycle is cooling. The rate of cooling to be applied for the austenite transformation depends on the heat treatment's purpose and the kind of change in mechanical properties desired. Since there is a strong relationship between microstructure and properties of steel hence the kind of microstructure a component should have must be ascertained first in light of desired mechanical properties, e.g. for high hardness only martensitic structure, for softness coarse-grained pearlitic structure, ferrite and for increasing the strength, hardness and toughness fine pearlite or bainite should be present. The controlled cooling rate can transform the austenite into all these phases and can result in a wide variety of mechanical properties in a given steel component. Therefore, each type of heat treatment uses a unique heat treatment cycle involving heating, soaking and cooling. Establishing a suitable heat regime, soaking time and the cooling condition is termed the design of heat treatment.

5.11 Common heat treatments of Steel

5.11.1 Annealing

The main feature of annealing heat treatment is the extremely slow cooling of steel after achieving the homogeneous austenitic state. The slow cooling transforms the austenite into coarse ferrite and pearlite. This microstructure helps soften the metals, increase the ductility and homogenise the composition and structure. Additionally, high-temperature exposure relieves the locked-in strain and residual stress, if any. However, it eliminates previous work/strain hardening effects due to recovery and recrystallization. According to the purpose, annealing is performed in the following ways:

- Box/Furnace Annealing: To soften the steel
- In-Process Annealing: To induce ductility for further mechanical working / processing
- Stress Relieving: To relieve residual stress

The following three basic steps are followed for annealing.

Heating: Box and process annealing need heating of steel to get a homogeneous austenitic state. The heating temperature depends upon the alloy composition, as mentioned above, and this is selected using the UCT line in the Fe-C diagram. Hypoeutectoid steels are heated to 30^0C -50^0C above the upper critical temperature, whereas hypereutectoid steels are heated to 30^0C -50^0C above the LCT. Heating to high temperatures reduces the work hardening effect and makes steel samples available for further processing (in-process annealing).

Soaking: Soaking time is determined theoretically based on alloy composition and dimensions of the sample to be heat-treated to get a homogeneous austenitic state. There is a sequence/order in which the decomposition of various phases (pearlite, ferrite/cementite, alloy carbides and homogenisation) takes place. Therefore, soaking time should be selected depending upon the composition and expected phases in a given steel sample. Increasing the heating temperature reduces the soaking time, but there are limits. Otherwise, too high a temperature may cause many other problems, such as coarsening of austenite grain, decarburisation, oxidation and cracking due to high-temperature gradient etc. As a thumb rule, 1 hour for a 25 mm diameter sample and then 30 minutes for each extra 25mm diameter is sufficient for getting the desired results.

Cooling: Furnace cooling is generally used in annealing (keeping the sample inside the furnace after switching it off). Furnace cooling provides extremely low cooling rates (3-5^{0}C/sec), resulting in coarse ferrite and pearlitic structure (CCT diagram). This structure is responsible for softening and increasing the ductility of steel.

Stress relieving

Annealing to relieve the residual stresses involves heating below the lower critical temperature range of 200^{0}C -600^{0}C, depending upon the level of stresses, followed by air cooling.

Spheroidizing

The spheroidizing treatment is a variant of the annealing heat treatment. The spheroidizing is performed on medium and high carbon steel to improve the machinability by softening steel and transforming the thick layers of cementite (in pearlite) into the ferrite and spherical nodule of cementite. This is achieved through prolonged exposure of steel at temperatures around lower critical temperatures (650-750 C).

5.11.2 Normalizing

The main feature of normalising heat treatment is to refine the microstructure using a moderately high cooling rate after austenitization. The moderate cooling rate (still / forced air cooling as per steel composition and hardenability) transforms the austenite into fine ferrite, pearlite, and bainite. This microstructure helps to increase hardness, strength, and toughness without compromising ductility. Like annealing, normalising also relieves the residual stress besides eliminating previous work/strain hardening effect, if any, due to recovery and recrystallization. Therefore, normalising heat treatment is used to achieve the following:

- Homogenise the structure.
- Increase strength, hardness and toughness.
- Relieve residual stress.
- Refine grain structure.

Heating and Soaking

These two steps of the normalising operation are similar to annealing, as discussed above.

Cooling: Slightly higher cooling rate is used for normalising than the annealing. However, the maximum cooling rate for normalising should be lower than the critical cooling rate as obtained from CCT diagram of the given steel to avoid any hardening/embrittlement. Air-cooling is extensively used in normalising operation (still air or current-air). Air-cooling provides a bit high cooling rate (10-35^{0}C/sec), transforming austenite into fine ferrite, pearlitic, and bainite (CCT diagram). This structure is responsible for strengthening, hardening and increasing the toughness of steel.

5.11.3 Quenching

It is also known as hardening treatment. Quenching is primarily used to increase the hardness, strength and wear resistance of steel.

Heating and Soaking: These two steps of hardening heat treatment for eutectoid and hypoeutectoid steels are mainly similar to annealing and normalising, as discussed above. However, the heating of hypereutectoid steel is carried out in the inter-critical temperature range, i.e. between the LCT and UCT. Therefore, in this process, cementite does not dissolve into austenite completely. Dissolution of cementite is unnecessary as cementite is a very hard phase. Moreover, carbon associated with cementite does not go into the austenite solution due to heating in the inter-critical temperature range. Reduction in the actual carbon content in austenite increases the critical cooling rate, reducing steel's hardenability. Moreover, presence of undissolved carbide particles promotes grain refinement due to heterogeneous nucleation because these carbide particles act as nuclei.

Cooling: Very high cooling rate (150-300^{0}C/sec) is used in hardening treatment than normalising. The cooling rate needed for hardening/quenching depends on the alloy composition, austenite grain size and homogeneity of austenite. These parameters affect the critical cooling rate and so hardenability of steel. High cooling rates (depending upon the requirements) can be achieved by quenching the steel sample in air, oil bath, plain water or salt water or in a bath kept under sub-zero conditions. For complete hardening through the transformation of austenite into martensite, the applied cooling rate must be higher than the critical cooling rate. High cooling rate results in the martensitic structure (see CCT diagram). This structure is responsible for hardening Steel (Fig. 5.11). Amount of martensite formed as a result of austenite transformation depends on the quenching temperature. The temperature of cooling medium (quenchant) must be below the M_f temperature for complete transformation of austenite into the martensite.

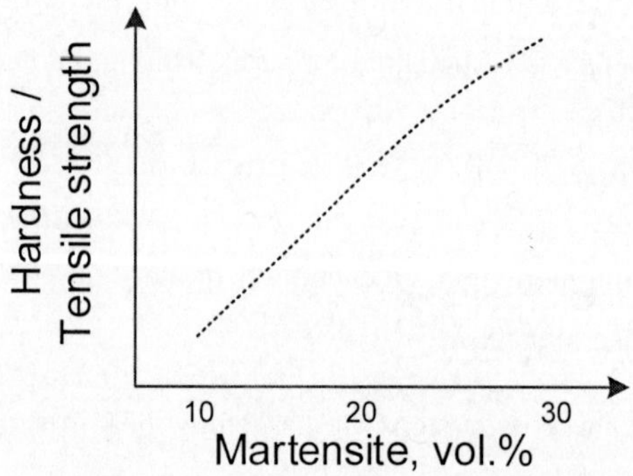

Figure 5.11: Schematic diagram showing the effect of martensite on hardness and tensile strength

Hardening Through the Section

In this case, cooling rate in the entire cross section is higher than the critical cooling rate required for martensitic transformation. On quenching from high temperature, surface of the sample attains the bath temperature immediately and causes martensitic transformation. Subsequently austenite to martensite transformation takes place across the section from the surface to core as evident from Figure 12.

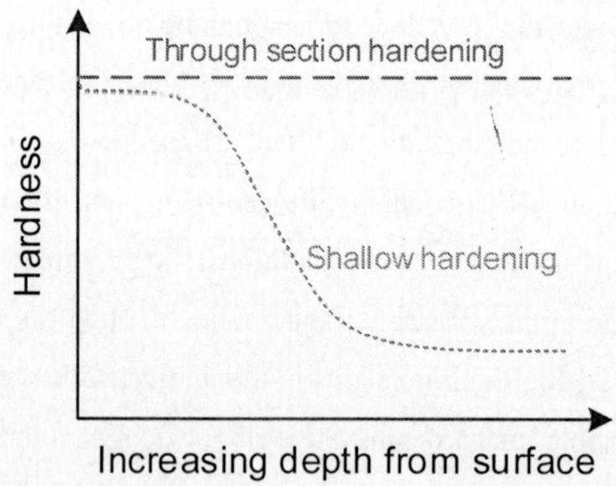

Figure 5.12: Schematic diagram showing hardness variation across the section of steel subjected to through section and shallow hardening

Change in specific volume due to martensitic transformation at the surface is accommodated by the deformation of a metal at the core as it is soft and ductile (austenite) at high temperatures. However, when the metal at the core is subjected to austenite to martensitic transformation, the increase in specific volume (expansion) is restricted by a hard and brittle layer of the martensite around the core (near the surface), as shown in Figure 5.13. This differential expansion and contraction at different stages leads to tensile residual stress at the surface, whereas the core experiences the compressive type of residual stress.

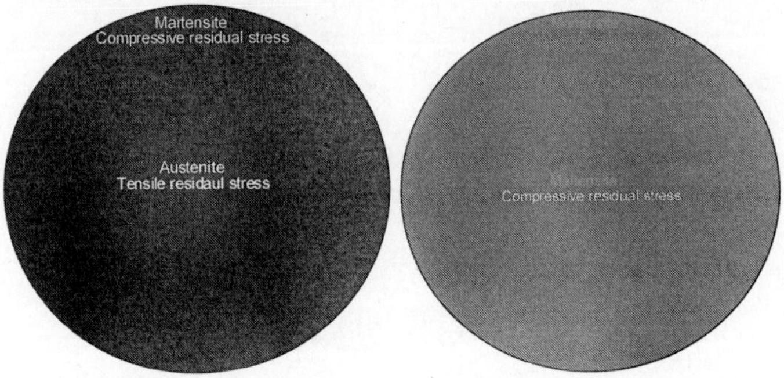

a) b)

Figure 5.13: Schematic diagram showing residual stress development during through section hardening treatment of steel when a) steel surface cooled to quenching temperature but the core is still hot, and b) hardening treatment is complete and sample cooled to room temperature

Shallow Hardening

Hardening treatment is termed as shallow hardening when austenite martensitic transformation causing an increase in hardness is localised at the near-surface layers only (Fig. 5.12) while at the core, soft phases like pearlite, ferrite, and bainite as per steel composition are formed. In shallow hardening, the cooling rate at the surface is higher than that required for martensitic transformation. Therefore, on quenching from high temperatures, the sample's surface attains the bath temperature immediately, while the core remains at a high temperature for a long time. This leads to a high-temperature gradient across the section. Therefore, during hardening treatment, steel experiences (1) a decrease in volume due to a reduction in temperature and (2) an increase in volume due to martensitic transformation.

The formation of martensite at the surface takes place by an increase in volume. This volume change is accommodated by the deformation of core metal (as the surface is already hard), which is soft and ductile at high temperatures than the surface layers. As the temperature of steel at the core decreases, shrinkage is restricted to the surface layer as they are hard, ductile and rigidly attached to the core metal. This differential expansion and contraction leads to the development of compressive residual stress at the surface, whereas the core experiences a tensile type of residual stress (Fig. 5.14).

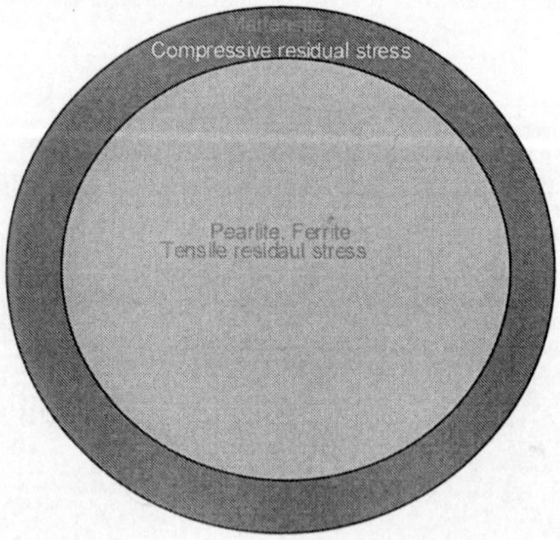

Figure 5.14: Schematic diagram showing residual stress development during shallow hardening treatment of steel

Quench Cracks: Austenite to martensite transformation takes place with a significant increase in specific volume. Therefore, all those regions (where martensitic transformation occurs) are subjected to compressive residual stresses. For the sake of equilibrium, all those regions where no such transformation takes place are subjected to tensile residual stresses. Residual stresses are those locked-in stresses present in steel, even if there is no external load. Compressive residual stresses are desirable as these resist the nucleation and propagation of cracks, improving fatigue strength/life. Tensile residual stresses adversely affect tensile properties (strength and ductility), fatigue strength, and fracture resistance, as these promote cracks' nucleation and propagation. During the hardening of the high carbon steels, development of high hardness coupled with low ductility leads to the formation of quench cracks in the presence of the tensile residual stresses; as the martensitic structure is extremely hard and brittle, which does not accommodate any yielding due to residual stresses which lead to cracking of hardening steel.

The nature of residual stresses present at the surface of quenched steel is very important as it decides whether these stresses are favourable or unfavourable. This nature depends on whether hardening is shallow or through the section. Shallow hardening develops compressive residual stress at the surface, whereas hardening through the section generates tensile-type residual stress. The presence of tensile residual stress on hardened steel's hard and brittle surface causes quench cracks (Fig. 5.15).

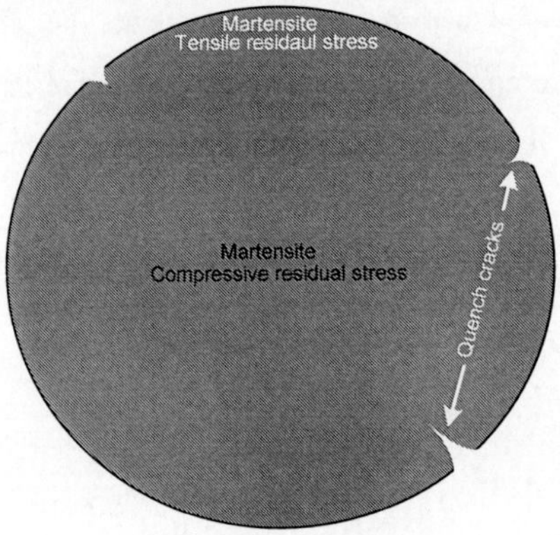

Figure 5.15: Schematic diagram showing quench cracks due to tensile residual stress caused by hardening treatment of Steel

5.11.4 Tempering

In general, all the hardened samples are tempered to relieve the residual stresses and induce some ductility and toughness at the cost of hardness. Hardened steel in as-quenched condition becomes extremely brittle and can not take impact load (toughness 5-10 J). Tempering of hardened Steel is done by heating the components in a range of temperature from 200^{0}C to 700^{0}C depending upon the degree of increase in toughness is acceptable at the cost of hardness. The heat treatment cycle used for tempering after quenching is shown in Fig. 5.16.

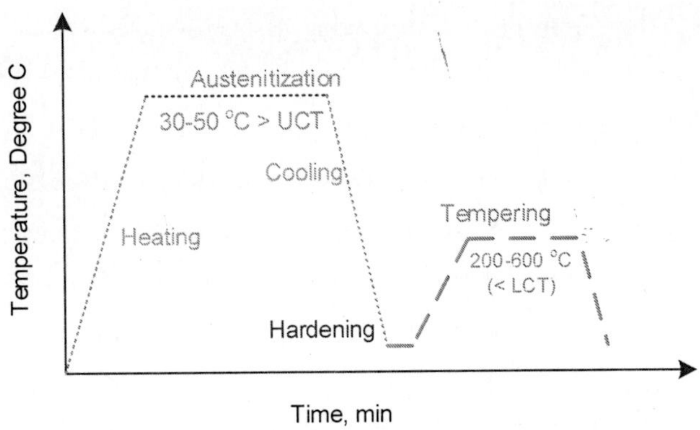

Figure 5.16: Schematic diagram showing the heat treatment cycle used for hardening followed by tempering steel

Variations in toughness and hardness with an increase in tempering temperature depend on steel composition (Fig. 5.17). If there are no carbide-forming elements (Cr, W, V), then there is a sharp reduction in hardness with an increase in tempering temperature in carbon steel. In the case of alloy steel, have few carbide forming elements, then little increase in hardness is noticed on tempering in the temperature range 500-600°C due to secondary hardening after initial reduction. Secondary hardening is due to the formation of hard alloy carbides due to the reaction between the carbide forming elements and carbon present in the steel. Martensite having entrapped carbon in interstitial spaces. Carbon diffuses out of unit cells (BCT) of hardened steel at high temperatures during the tempering.

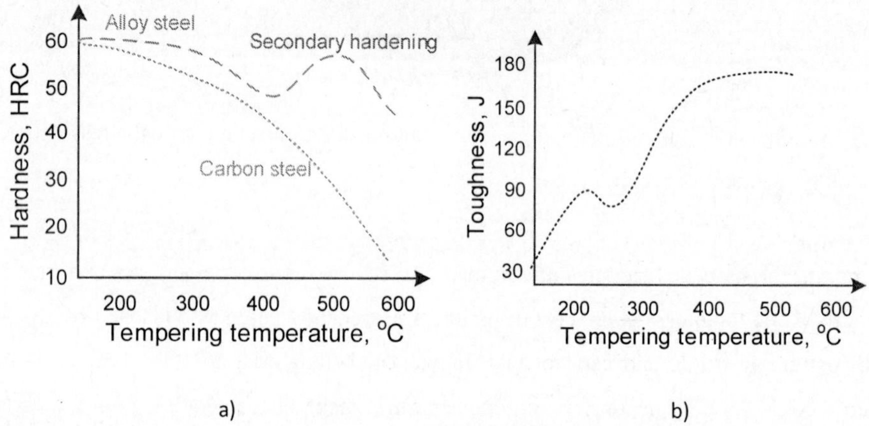

Figure 5.17: Effect on tempering temperature on a) hardness and b) toughness of hardened carbon and alloy steel

The diffusion of carbon from martensite (having a supersaturated carbon solution in Fe) reduces the degree of lattice distortion (c/a ratio). Rejected (diffused out) carbon reacts with iron to form cementite. When the carbon content reduces to the critical limit (corresponding to the solubility of carbon in α-Fe) martensite having B.C.T. lattice structure is converted into α-Fe (ferrite). Prolonged heating of hardened steel samples causes spheroidization of cementite in the matrix of ferrite. Increased ductility, in combination with reduced residual stress, decreases the cracking tendency. Formation of quench cracks can be reduced by using two methods of hardening discussed below:

5.11.5 Austempering

Austempering heat treatment involves the isothermal transformation of austenite into bainite. Steel is heated to get a homogeneous austenitic state, and then it is quenched in a bath kept at a temperature below the nose and above the martensite transformation start temperature (M_s) in CCT curve (Fig. 5.18). Sample is maintained in the bath until the whole of the austenite transforms into bainite. Holding the sample at a constant temperature equalises the temperature across the section, avoiding the differential volumetric change due to temperature difference and phase transformation. This reduces the residual stress, cracking and distortion tendency. Further, it does not involve any martensite formation in this treatment. The transformation of austenite into fine bainite increases the strength and hardness without a significant drop in toughness.

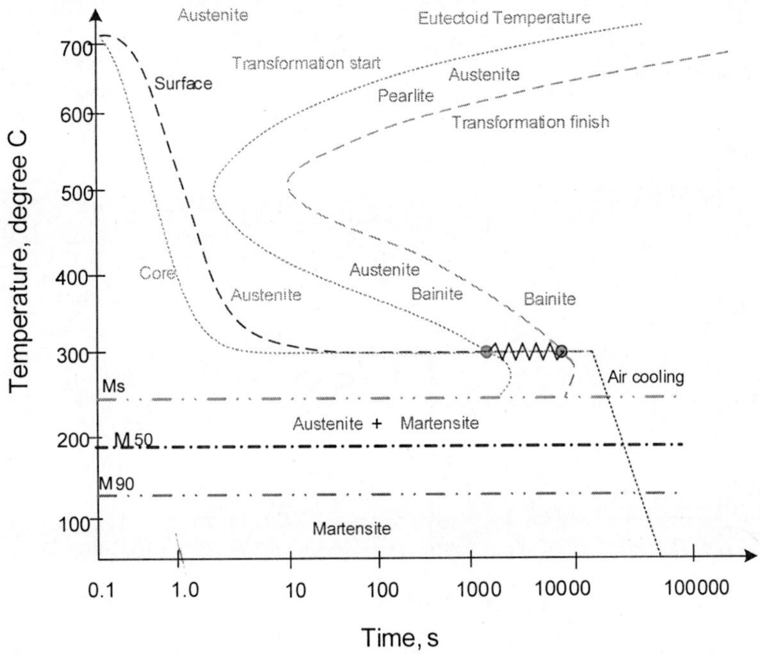

Figure 5.18: Schematically austempering treatment of eutectoid steel is shown with the help of CCT diagram

5.11.6 Martempering

The martempering heat treatment uses two-stage quenching for martensitic transformation, followed by hardened steel tempering to achieve the desired combination of strength and toughness (Fig. 5.19). steel heated to get homogeneous austenitic state is quenched in a bath kept at a temperature just above the martensite transformation start temperature (M_s) in CCT curve. Steel is kept in the bath until the temperature of the whole cross-section becomes uniform. Then the steel is again quenched in a bath maintained at a martensite transformation finish temperature (M_f). The uniformity of the temperature across the section reduces the temperature gradient during hardening treatment (second quenching), which ensures simultaneous transformation of austenite into martensite in the entire cross section. Thereby reducing the development of residual stresses. The absence of residual stresses avoids quench cracks on the hardened sample. Then tempering is done.

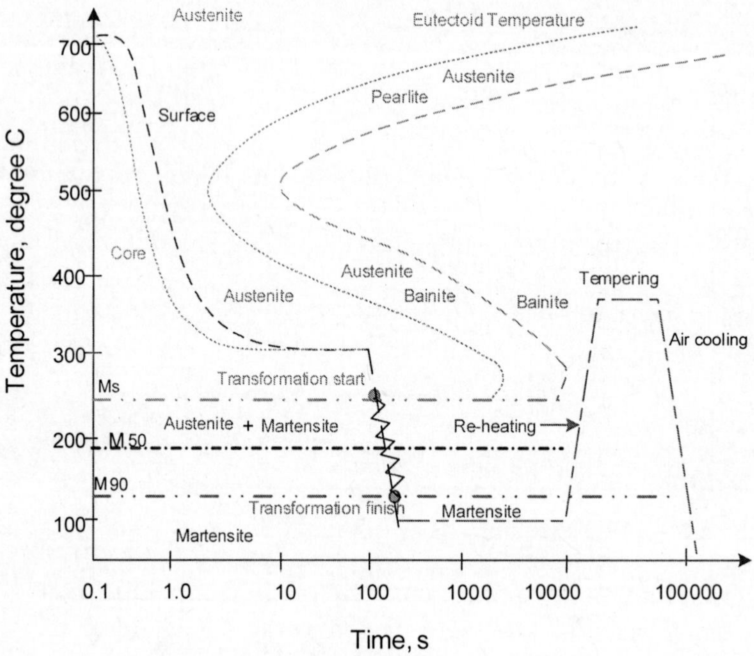

Figure 5.19: Schematically martempering treatment of eutectoid steel is shown with the help of CCT diagram

5.12 Case hardening

Case hardening primarily involves increasing the surface hardness for improved wear and fatigue resistance. Case hardening, however, is expected not to affect the subsurface metal properties (Fig. 5.20). Thus, case hardening helps to develop components with hard surfaces and tough core / sub-surface.

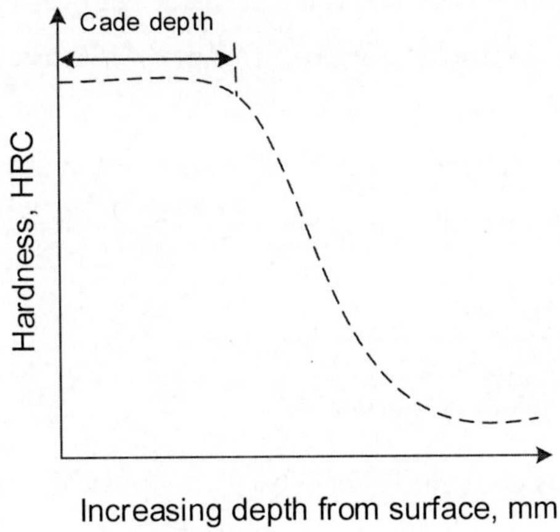

Figure 5.20: Schematic diagram showing variation in hardness of case hardened component

5.12.1 Need of case hardening

Many engineering applications need a combination of two opposite kinds of properties, such as toughness and hardness. For example, the surface of a tribological component should be hard for wear resistance, whereas core should be tough to withstand under the impact loads. Such a dissimilar combination of mechanical properties is achieved by case hardening. In the case of plain carbon steels, an increase in carbon content increases the hardness but at the cost of toughness. To overcome this problem, two different kinds of case hardening approaches are used for a) steels having carbon less than 0.3% C and b) steels having carbon more than 0.5%. A combination of properties hardness/toughness, can only be obtained by controlling the microstructure suitably across the component section. As we know that the, high hardness in steels can be obtained by developing the martensitic structure, whereas, for increased toughness, steel should have a combination of soft and relatively hard phases such as ferrite and pearlite bainite.

Case hardening helps to achieve the hard constituents like martensite, nitrides and carbides at the surface while retaining tough phases in the core. It is easier to obtain the martensite in high carbon steel (C> 0.8%) than the low carbon steel (C< 0.3%). Hence, case hardening of the low carbon steel needs the first addition of carbon so that martensite of high hardness can be produced without difficulty during subsequent hardening treatment. Shallow hardening is also an example of the case hardening of high carbon steel without the addition of carbon.

5.12.2 Principle of case hardening

Improvement in surface properties through case hardening can be achieved using three approaches a) surface modification (without change in chemical composition) through mechanical and surface treatment, b) surface modification via the change in the chemical composition of surfaces by introducing suitable alloying elements and c) building up a layer of suitable material on the surface of the component to achieve the desired surface characteristics. The first two approaches will be presented in this unit, and the third approach, based on coating, cladding, and reclamation, is not in the scope of this subject.

5.12.3 Physical metallurgy of case hardening

Case hardening techniques (with or without change in chemical composition) are primarily based on achieving the desired microstructure using either thermo-mechanical treatment or changing the chemistry of the base metal surface, also called the substrate. Case hardening techniques (without change in chemical composition) rely on metal strengthening mechanisms like strain hardening, grain refinement, and transformation hardening. The controlled surface layer deformation is used to harden the surfaces (through grain refinement, strain hardening, and even transformation hardening in a few cases like steel and copper) using techniques like shot peening, burnishing, and friction stir processing. Thermal treatments like laser hardening, induction hardening and flame hardening are based on the transformation hardening (austenite to martensite) using controlling the heating and cooling cycle.

The case hardening by changing the chemical composition of the surfaces is based on the simple principle of developing suitable microstructure through controlled alloying followed by suitable heat treatment. Steel surfaces are frequently hardened by carburising, nitriding, carbo-nitriding, and cyaniding. In the case of carburising, carbon is introduced at the surface of low Carbon Steel (< 0.3% C). Nitriding, on the other hand, introduces nitrogen at the surface of the steel. Cyaniding and carbo-nitriding introduce carbon and nitrogen at the surface of the Steel component to be case hardened.

The following section presents the general approach, heat treatment cycle, and the related parameters of various case hardening processes

5.12.4 Case Hardening Operation without Change in Composition

Case hardening methods of this category are mainly used to harden the surface of high carbon steel. These methods are based on the principle of shallow hardening. Shallow hardening relies on martensitic transformation by controlled heating to achieve the austenitic state, followed by rapid cooling to get a martensitic structure. Depending upon the methods of heating adopted for case hardening, the following are a few common methods of this group:

- Flame hardening
- Laser hardening
- Induction Hardening

5.12.4.1 Flame Hardening

This method uses oxy-acetylene flame for heating the surface of medium and high carbon steel (to be hardened) followed by rapid cooling (quenching) by a water jet moving along with the heating gas torch (Fig. 5.21). Process parameters like travel speed of torch, distance between torch and water, torch tip to surface distance all affect the heating and cooling cycle which in turn determine the depth of the hardened case.

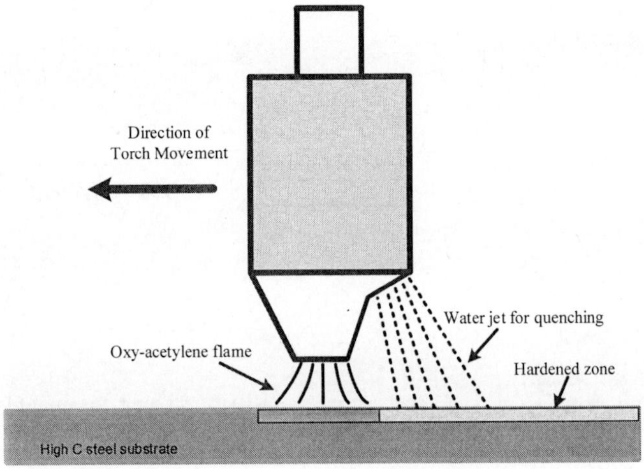

Figure 5.21: Schematic diagram showing the arrangement of flame hardening of high carbon steel

5.12.4.2 Induction hardening

This method is based on joule (electrical resistance) heating the principle by inducing the current at the surface layer (of the component to be hardened) using high frequency current passed through an induction coil. The induction effect achieved by bringing a coil (through which high frequency current passed) close to the surface. Depending up on the frequency of current through the coil, eddy current is induced up to a certain depth in the steel to be hardened, which causes heating (by electrical resistance heating) of the material in very short time. The heated material is then quenched for necessary martensitic transformation to harden the surface.

Quenching can be done either by the jet of water or just by switching off the induction heating, so the material below the surface is self-quenched due to rapid cooling by the substrate itself. In this process, hardened case depth mainly depends on the current frequency passed through the induction coil. Case depth and frequency are inversely related. High frequency current results in shallow case depth (Fig. 5.22).

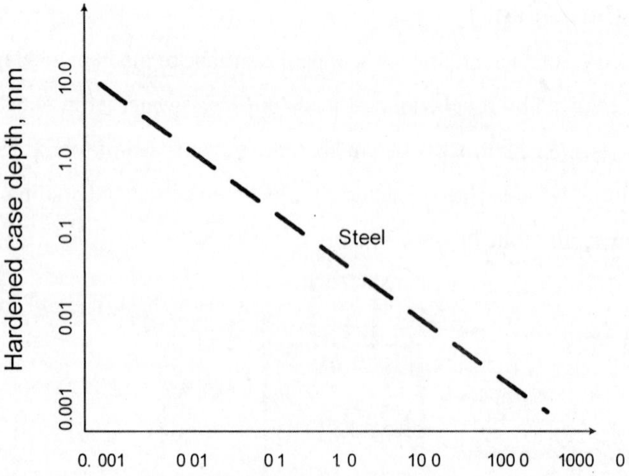

Figure 5.22: Effect of frequency of current used for induction hardening on case depth

5.12.4.3 Laser Hardening

The laser hardening process uses a laser (light amplification and simulated emission of radiation) beam to provide a pulse high power density over a localized area per beam diameter. The laser beam causes fast heating and austenitization of the steel surface layer. As the beam moves away from the heated zone self-quenching by substrate causes rapid cooling to facilitate austenite to martensite transformation to produce hardened case. The laser scanning speed, bean diameter, and laser power are a few parameters, which need to be optimized for achieving the desired case depth (Fig. 5.23).

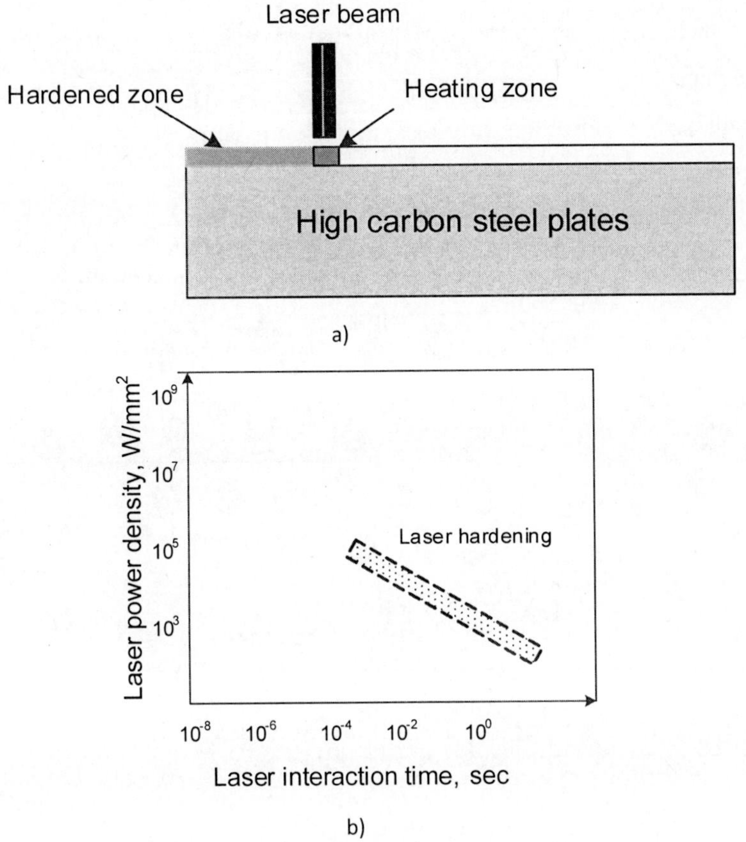

Figure 5.23: Schematic diagram showing a) scheme of laser hardening of high carbon steel and b) laser power & interaction time combination desired for laser hardening

5.12.5 Case Hardening by Changing the Surface Composition

Methods of this category are mainly used for case hardening of low carbon steel. These are based on the principle of changing the chemical compositions (using suitable elements) of the surface and near surface layer, followed by heat treatment to get desired microstructure (Fig. 5.24). However, the concentration of alloying element decreases with increasing depth from the surface which in turn results in varying phases, compound and mechanical properties. As per the requirement of case depth, alloying concentration must be increased up to the minimum acceptable and desired level using a suitable combination of temperature and time surface treatment during case hardening. The addition of carbon facilitates the formation of a hard martensitic structure on subsequent heat treatment. Martensitic transformation at the surface is obtained by controlled heating to the austenitic state followed by rapid cooling. Nitriding and carbo-nitriding methods introduce nitrogen into near surface layers of steel (carbon and alloy steel) to form hard and stable nitrides. Methods of this category are listed below:

- Carburizing
- Cyaniding
- Nitriding
- Carbo-nitriding

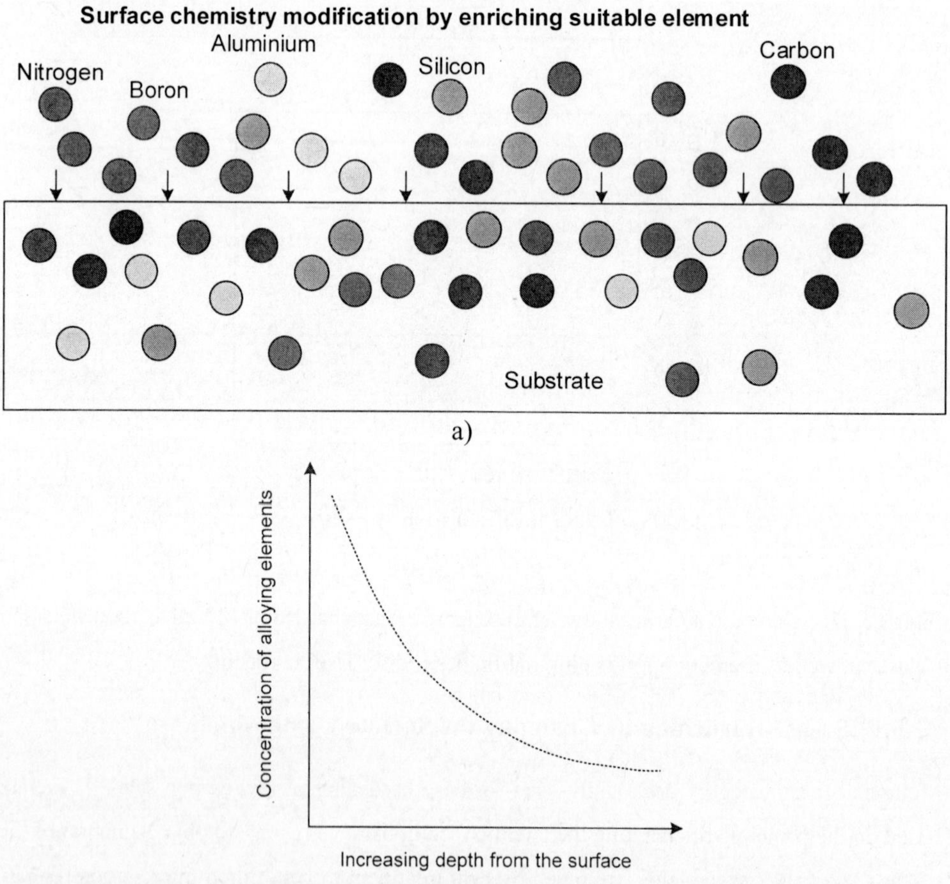

Figure 5.24: Schematic diagram showing a) case hardening / surface engineering approach by modifying the surface chemistry and b) variation in concentration of the alloying elements with increasing depth from the surface

5.12.5.1 Carburizing

In this method, carbon content near the surface layers of the low carbon steel component is increased by putting the component in an active carbon-rich environment at a high temperature. Heating of steel component to the carburizing temperature range (850-950 °C) transforms ferrite and pearlite of steel into the austenite. At high temperatures, austenite offers high solid solubility to the carbon. High temperature also increases diffusion rate, reducing the time required for carbon enrichment up to the desired depth. According to the kind of medium being used for carburizing, the process is termed as solid, liquid and gas carburizing. Carburizing increases the carbon content of the steel surface layer only, subsequent heat treatment is necessary for producing hard martensite and realize desired case hardening (Fig. 5.25).

Solid carburising

This carburising process uses a mixture of charcoal, and barium carbonate (energiser) in a heated chamber. The component to be carburised is packed in a closed chamber along with the above mixture. The combustion of charcoal in the presence of limited oxygen produces carbon dioxide and carbon mono-oxide, which in a chemical reaction produces active carbon. The active carbon in an atomic state at high temperature diffuses into the steel surface. However, solid carburising suffers from many limitations, such as poor control over the temperature and the reactions inside the chamber. Direct quenching of carburised components is not possible. Poor control over the temperature and reactions inside the chamber makes it difficult to control the desired case depth. Since direct quenching is not possible in pack carburising, reheating the component for hardening treatment is needed.

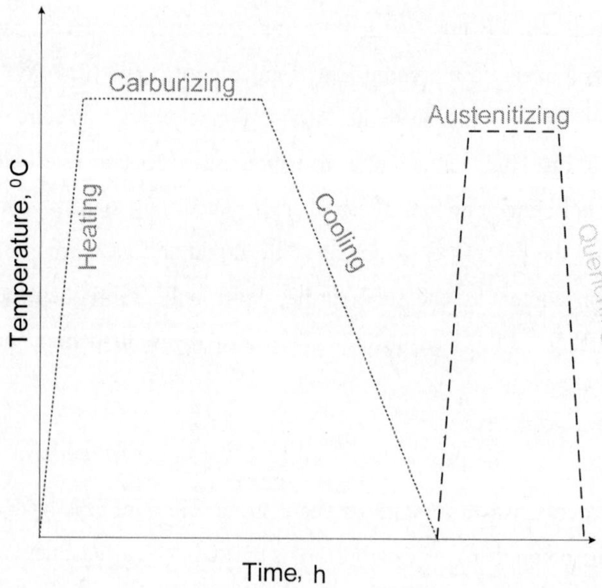

Figure 5.25: Schematic diagram showing the thermal cycle of solid / pack carburising treatment for case hardening

Liquid carburising

The liquid carburising uses a molten bath having a mixture of sodium cyanide, and sodium carbonate. Liquid bath causes rapid heating due to good thermal contact of the components with high temperature liquid, which significantly reduces the carburising time. This process overcomes both the problems of solid carburising because in this method component after carburising can be quenched directly from the austenitic state for case hardening (Fig. 5.26). Further, temperature of the bath and so chemical reactions can be easily controlled. A good control over heating temperature and time makes it possible to have closer control over the case depth.

$BaCl_2 + 2\,NaCN \rightarrow Ba(CN)_2 + 2NaCl$

$Ba(CN)_2 + Fe \rightarrow Fe(C) + BaCN_2$

A few additional reactions may also occur to introduce nitrogen in Steel:

$NaCN + O_2 \rightarrow 2NaCNO$

$3NaCNO \rightarrow Na_2CO_3 + NaCN + C + 2N$

Gas carburising

The gas carburising uses a gaseous mixture of propane and butane, which decomposes at high temperatures to produce nascent atomic carbon. As per the concentration of carbon atoms in the gaseous mixture, its flow rate needs to be closely controlled. In case of a high concentration of carbon atoms in the mixture, the flow rate should be low enough to avoid the deposition of soot (deposition of a carbon layer) on the steel surface. The flow rate of the gaseous mixture should be such that carbon atoms available at the surface for penetration must be equal to those which are diffusion into the surface. Like liquid carburising, gas carburising also allows direct quenching (after carburising) from the austenitic state for case hardening.

$C_3H_8 \quad 2CH_4 + C$

$CH_4 + Fe \quad Fe(C) + 2H_2$

$CH_4 + CO_2 \quad 2CO + 2H_2$

$2CO + Fe \quad Fe(C) + CO_2$

$H_2 + CO_2 \quad CO + H_2O$

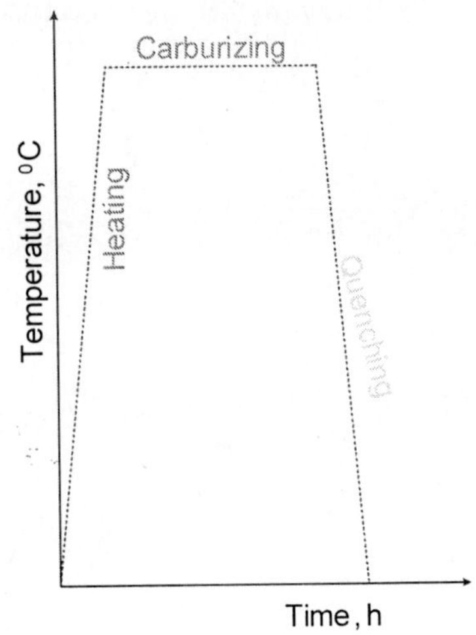

Figure 5.26: Schematic diagram showing the thermal cycle of liquid/gas carburising treatment for case hardening

5.12.5.2 Cyaniding

The cyaniding is similar to that of liquid carburising except that the mixture of sodium cyanide, sodium carbonate and sodium chloride has a higher concentration (90%) of sodium cyanide. In this process, case hardening mainly occurs due to the diffusion of nitrogen and the formation of iron nitride and alloy nitride. Penetration of carbon does not contribute much in case of hardening by this method, while the reverse is true for carbo-nitriding. Heat treatment is not carried out after cyaniding. Case hardening in cyaniding is due to nitride formation, whereas in carburising, it is due to martensite formation. The following reaction takes place during the cyaniding.

$$2NaCN + O_2 \rightarrow 2NaCNO$$

$$2NaCNO + O_2 \rightarrow Na_2CO_3 + CO + N_2$$

$$2CO \rightarrow CO_2 + C$$

5.12.5.3 Nitriding

Nitriding of Steel is performed at lower temperatures than carburising. Nitriding diffuses nitrogen into the carbon and alloy steel surface and forms hard nitrides near the surface layer for case hardening (Fig. 5.27). Nitriding of Steel is carried out in the ferritic zone at about 500^{0}C -550^{0}C so as to achieve a reasonably high diffusion rate for penetration of nitrogen without transformation of ferrite and pearlite into austenite. Nitrogen reacts with iron and nitride forming element like Cr, Al V to form their iron nitrides. Nitrogen rich environment is generated by ammonia through the following reaction.

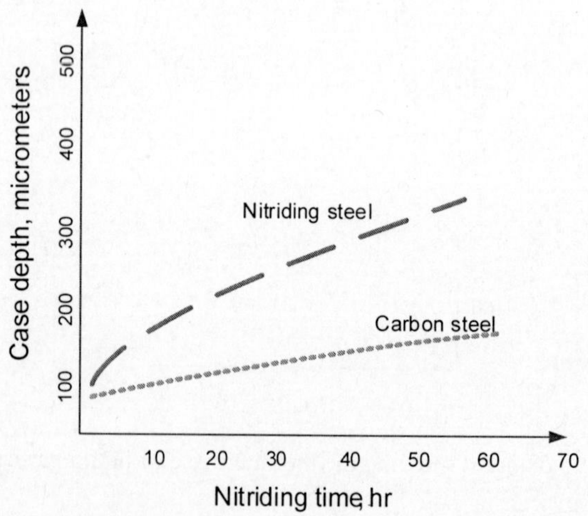

Figure 5.27: Effect of nitriding time and steel composition on the depth of case hardening

To increase effectiveness of this method a few strong and stable nitride forming alloying elements such aluminium, vanadium and chromium are added in steel such steels are called nitriding steels (Fig. 5.27). However, the concentration of nitrogen gradually decreases with increase of the depth from the surface. Varying nitrogen concentration results in different nitrides near the surface layers (Fig. 5.28). Formation of thick iron nitride layer near the surface layers (appears white under a microscope) therefore, it is called the white layer. The white layer becomes hard, brittle and crack sensitive. These layers are chipped off during the actual use of hardened surfaces. Therefore, white layer must be removed before putting the component in use.

$$2NH_3 \longrightarrow 2N + 3H_2$$

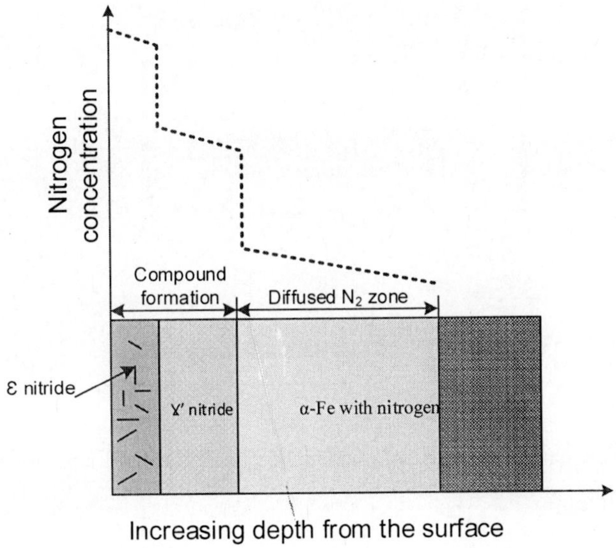

Figure 5.28: Schematic diagram showing varying nitrogen concentration with increasing depth and nitrogen compound formation due to nitriding for case hardening of Steel

5.12.5.4 Plasma Hardening

The plasma hardening can be done using plasma carburising or nitriding following the same basic principle of enriching steel surface with carbon and nitrogen, respectively, of conventional carburising and nitriding processes. However, the plasma hardening process uses active carbon and nitrogen atoms obtained directly through plasma formation. The plasma is generated using a suitable temperature (300 to 800 ^{0}C) and pressure (300 to 1000 Pa) instead of relying on the thermal decomposition of hydro-carbon and ammonia (Fig. 5.29). This feature makes plasma hardening processes faster than conventional carburising and nitriding processes. This process is also called glow discharge plasma, as a glowing light is found around the work piece plasma, carburising and nitriding. The formation of carbide and nitrides at the near surface layers increases the hardness, wear and fatigue resistance of the steel components.

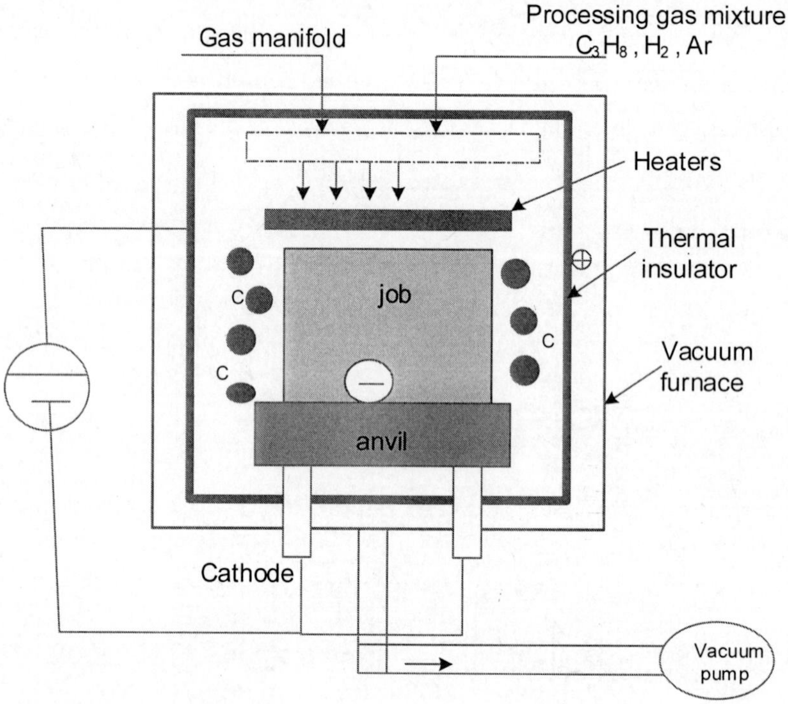

Figure: 5.29 Schematic diagram showing the general approach of plasma hardening (carburising/nitriding)

UNIT SUMMARY

This unit presents the fundamentals of modifying the microstructure of the steel using suitable heat treatment cycles. Selection of the heating temperature for austenitization, followed by soaking time and control cooling, is crucial for developing the desired microstructure and Mechanical properties during the heat treatment of steel. The factors affecting the heating temperature, soaking time, and cooling rate have been described using suitable schematic diagrams. The heat treatment cycles to be used for the common heat treatments like annealing, normalising, quenching, tempering, austempering, and martempering have been presented. Additionally, the fundamentals and need of the Steels case hardening and common case hardening methods like carburising; nitriding, cyaniding, and plasma hardening have also been described.

EXERCISE

Questions for self-assessment

1. The heat treatment used to induce toughness in hardened steel is
    a. Annealing
    b. Normalising
    c. Spheroidizing
    d. Tempering
2. Annealing heat treatment of carbon steel is done at
    a. Below recrystallization temperature
    b. Below lower critical temperature
    c. Above upper critical temperature
    d. Above eutectic temperature
3. The heat treatment which uses maximum cooling rate is
    a. Annealing
    b. Hardening
    c. Normalising
    d. Tempering
4. Heat treatment of steel designed to produce bainitic structure only is
    a. Annealing
    b. Austempering
    c. Martempering
    d. Tempering
5. Case hardening treatment of steel at the surface results in maximum
    a. Hardness
    b. Toughness
    c. Ductility
    d. All of these
6. Heat treatment which combines both hardening and tempering, is
    a. Hardening
    b. Austempering

c. Martempering
d. Tempering

7. The heat treatment of steel performed at minimum temperature is
    a. Spheroidizing
    b. Normalising
    c. Martempering
    d. Tempering
8. The transformation of austenite into various phases at constant temperature is shown by
    a. Fe-C diagram
    b. TTT diagram
    c. CCT Diagram
    d. All of these
9. Only carbon enrichment of low carbon steel surface for increasing the surface hardness achieved during
    a. Carbo-nitriding
    b. Cyaniding
    c. Carburizing
    d. All of these
10. The lowest critical cooling rate is shown offered by
    a. Hypoeutectoid Steel
    b. Eutectoid Steel
    c. Hypereutectoid steel
    d. All of these
11. The heat treatment for which the carbon equivalent should be high is
    a. Annealing
    b. Normalising
    c. Tempering
    d. Hardening
12. White layer on the surface of alloy steel (Fe-C-Cr-V-Al) is formed during
    a. Carburizing
    b. Nitriding
    c. Hardening
    d. Laser hardening

13. The residual stress in a rolled component of steel is minimised after
    a. Shallow hardening
    b. Through section hardening
    c. Normalising
    d. All of these
14. The hardest phase formed after hardening heat treatment of a medium carbon steel is
    a. Bainite
    b. Pearlite
    c. Martensite
    d. Ledeburite
15. The most important factor of induction hardening affecting the case depth is
    a. Melting temperature of steel
    b. Thermal diffusivity of steel
    c. Coil to substrate distance
    d. Frequency of current supplied to induction coil

Answers of Multiple Choice Questions

Key for MCQ: 1 d, 2 c, 3 b, 4 b, 5 a, 6 c, 7 d, 8 b, 9 c, 10 c, 11d, 12 b, 13 c, 14 c, 15 d

Short and Long Answer Type Questions

1. Why does carbon steel allow change in mechanical properties with heat treatment?
2. Using Fe-C diagram enlist phase transformations which will on heating 0.2% steel to 1100 °C.
3. What is the inter-critical temperature for eutectoid steel?
4. How does the inter-critical temperature range affect with an increase of C%?
5. How does the carbon wt.% affect the heat temperature for the annealing of carbon steel?
6. Describe the effect of quenching temperature on isothermal transformation phase transformation between lower critical (LCT) and martensite start (Ms) temperature.
7. Why does TTT diagram takes a typical C shape?
8. How does the composition of steel affect CCT diagram?
9. Describe the relevance of the CCT and TTT diagram in hardening and austempering heat treatment.
10. Why do we need case hardening?

11. What is the fundamental approach of case hardening by laser and induction hardening?
12. Why does the carburising performed a) in low-carbon steel only and b) in the austenitic state of carbon steel?
13. How does the overheating temperature (above UCT) deteriorate the mechanical properties of steel after heat treatment, like normalising?
14. Why does the normalising of a low carbon steel produces a finer grain structure than annealing heat treatment?
15. Describe the residual stress developed in through-section hardened and shallow hardened carbon steel?

Numerical Problems

- Using the Fe-C diagram, the CCT diagram design annealing heat treatment cycle 0.1 % and 0.8% C steel
- Determine the carbon equivalent and Ms temperature of the following steels
  - Fe-0.3%C-2.25%Cr-0.5%Si-0.8%Mn-1.0%Mo
  - Fe-0.6%C-0.5%Si-0.8%Mn

PRACTICAL

1. Polish, etch and study the microstructure of as received rolled carbon steel sample and compare it with annealed, normalised, hardened samples of the same steel using optical microscopy.
2. Perform annealing, normalising, and hardening heat treatment of carbon steel (say AISI 1020) and study the effect of the above heat treatment on the hardness, and tensile properties of the same steel.
3. Perform tempering of hardened carbon steel (say AISI 1020) at an increasing temperature from 200 to 600°C and study the effect of tempering temperature on hardness and impact toughness using the Izod / Charpy test.
4. Using image analysis (like Image J open source software), determine the grain size and % of pearlite and ferrite from the microstructure of carbon steel (say AISI 1020) in annealed, and normalised conditions
5. Perform hardening heat treatment of carbon steel (say AISI 1020) on increasing section size, say 10, 20, 30 mm and then measure hardness variation across the section.
6. Design a hardening and annealing heat treatment cycle for carbon steel 0.4% C steel (say AISI 1040) and then study the feasibility of the same.

KNOW MORE

Take the steel samples and study their microstructure and mechanical properties. Based on the microstructure and hardness observations, estimate heat treatment, if any, that would have been performed. Conduct the toughness test on such samples and comment on ductility and impact resistance. Compare the structure and properties of the steel samples subjected to varying heat treatment. Take real world components like case hardened gear tooth, and machine tool functional surfaces. Prepare their cross-section and study hardness and structural variation with increasing depth from the surface.

REFERENCES AND SUGGESTED READINGS

1. Callister, Materials science and engineering, Wiley, Publication (2014)
2. S D Avner, Introduction to Physical Metallurgy, McGraw Hill, Publication (2002)
3. GE Dieter, Mechanical metallurgy, McGraw Hill, Publication (1989)
4. D K Dwivedi, Fundamentals of Metal Joining, Springer nature, Singapore (2022)
5. D K Dwivedi, Surface engineering, Springer nature, Germany (2018)
6. Ratna Kumar, NPTEL Course "Basics of Materials Engineering" https://onlinecourses.nptel.ac.in/noc20_me78/preview
7. D K Dwivedi, NPTEL Course, Fundamentals of Manufacturing Processes, https://archive.nptel.ac.in/courses/112/107/112107219/
8. Structure/Property Relationships in Irons and Steels, ASM International, Second Edition J.R. Davis, Editor, (1998), 153-173

Dynamic QR Code for the Unit 5

Unit 6

Common Engineering Metals: Mechanical Properties and Manufacturing

Alloy Steel, Stainless Steel, Cast iron, Alloys of Aluminium, Nickel, Copper, and Titanium

Unit Specific / Learning Objective

Objective this unit is to talk about following aspects

- To understand the importance of the various common metals
- To provide information about common alloys their compositions, mechanical properties
- To develop an understanding of the structure-property relationship for the different metals
- To learn the about ease of manufacturing using common metals
- To enlist application of common metals to make components

Additionally, few fundamental questions for self-assessment based on fundamentals **are** included in this chapter in form of recall, application, comprehension, analysis and synthesis. There are further suggested readings and reference for deep learners and readers assistance.

Rationale

Metals are widely used for making both consumer and capital goods. These goods are expected to perform under a wide range of service conditions ranging from the dry room temperature environment, static and fatigue loading, high-temperature oxidative environment, corrosive environment, salty environment, sub-zero low-temperature conditions, etc. Therefore, a single metal does not suit all kinds of service conditions for satisfactory performance and desired life. It is important to consider a certain set of the mechanical, tribological, and corrosive properties of the potential metals, which can be selected for making a component. Precisely the selection of suitable metal is based on the expected failure mechanism such as elastic deformation, plastic deformation, fracture due to overloading, fatigue, creep, impact, low-temperature, and high-temperature conditions, and corrosion. In addition to the properties, ease of manufacturing, the availability of infrastructure and resources, and finally economics certainly determines the selection of metal. Therefore, it is important to be aware of the different metals and their alloys available for selection to make the components. This unit provides detailed information about common metals (Fe, Al, Cu, Ni, Ti) their alloys, mechanical properties and their structure property relationship.

Pre-Requisites

A course on chemistry: Class XII

Learning outcomes

U6-O1: Ability to differentiate the various types of ferrous metals, and their suitability based on their mechanical properties.

U6-O2: Ability to relate the microstructure and mechanical properties of cast irons

U6-O3: Ability to understand the capabilities of Cu, Al, Ti, Ni, alloys based on mechanical, and corrosion properties

U6-O4: Ability to choose suitable metals for making components in light of the service conditions

U6-O5: Ability to select the suitable manufacturing process based on physical properties and metallurgical aspects

Unit-4 Outcomes	*EXPECTED MAPPING WITH COURSE OUTCOMES (1- Weak Correlation; 2- Medium correlation; 3- Strong Correlation)*					
	CO-1	*CO-2*	*CO-3*	*CO-4*	*CO-5*	*CO-6*
U6-O1	1	3	2			
U6-O2	1	3	2			
U6-O3	1	3	2			
U6-O4	1	3	2			
U6-O5	1	3	2			

Course Objective

1. Student will be able to identify crystal structures for various materials and one could understand the defects in such structures.
2. Understand how to tailor material properties of ferrous and non-ferrous alloys?
3. How to quantify mechanical integrity and failure in materials?

6.1 Introduction

Metals play an essential role in developing components and systems to be used by society. Apart from manufacturing, metals significantly determine the component's performance and reliability. The choice of suitable metal for manufacturing a particular component largely depends on the service conditions, especially with regard to the load, temperature, and environment. Accordingly, suitable metal with the desired physical, mechanical, and tribological properties is selected. The metals in pure form show so very poor Mechanical properties like yield strength, hardness, toughness, and ability to withstand high temperatures. Adding suitable alloying elements in pure metals improves the mechanical physical and corrosion properties, making them more suitable for industrial exploitation. In this unit, metals of commercial importance like steel and cast iron, copper alloys, brass, bronze, aluminium alloys, nickel alloys, and titanium alloys will be described concerning their physical metallurgy, mechanical properties, and common approaches for manufacturing.

6.2 Ferrous Metals

There are three common ferrous metals, which are extensively used for fabricating various engineering components. These are wrought iron, cast iron, and steel. This classification is largely based on the carbon content (wrought iron < 0.1 %C, steel 0.1 to 2.0 % C, and cast iron 2.0 to 6.67% C), while other elements like Mn, Si, S, and P are found as traces. The effect of these elements (considered as traces and impurities) on the mechanical properties is negligible. However, efforts are always made to minimize the concentration of these impurities to produce clean ferrous metals. The concentration of sulphur and phosphorus is kept below 0.05%. The higher concentration of sulphur and phosphorus degrades the mechanical properties and ease of manufacturing significantly. Sulphur increases the hot tearing and solidification-cracking tendency during fusion welding, while sulphur addition improves the machinability of steel. The phosphorus makes the steel brittle, which in turn increases the cracking tendency. Other elements, namely manganese and silicon, form a solid solution with iron and increase the strength and hardness of steel.

6.3 Wrought iron

The wrought iron has a very low carbon content (<0.1%C) besides slag (2-3%) in the form of elongated inclusions. Due to the low carbon content, wrought iron offers very high ductility (coupled with low strength) and ease of manufacturing by bulk deformation-based processes. The slag present in wrought iron acts as a flux, which plays an important role in fusion welding to produce a clean weld. Low carbon content coupled with the slag results in the high weldability of wrought iron. However, elongated inclusions in wrought iron significantly reduce the Z-direction ductility, increasing the under-bead cracking and lamellar tearing tendency.

6.4 Steel

Steel (based on composition) can be grouped into two broad categories a) plain carbon steel and b) alloy Steel. Plain carbon steel primarily contains a controlled amount of carbon while other elements (Mn, Si, P, S) remain as impurities. Plain carbon steel is further categorized as low (<).25%), medium (0.25 to 0.5%C) and high carbon steel (>0.5%). An increase in carbon content increases the tensile strength, hardness, and wear resistance of plain carbon steel while toughness and ductility are adversely affected.

In addition to the carbon, the alloy steels have a controlled amount of other alloying elements like chromium, molybdenum, vanadium, tungsten, aluminium, etc. The addition of the alloying elements in steel, affects the physical metallurgy, microstructure (phase and grain structure) and mechanical properties (Table 6.1). Based on the total amount of alloying elements in alloy steel is grouped into three general categories, namely micro-alloy steel (alloying element < 1.0%), low alloy steel (1-10%) and high alloy steel (>10%). Further, many steels have been developed in each of the categories. Micro-alloy steel and low-alloy steel are designed for structural applications.

The most common high alloy steel is stainless steel with a minimum of 12.0 % Cr. The presence of Cr in steel resists general atmospheric, which makes it stainless. Conversely, stainless steel does not rust or decolorize when exposed to ambient conditions during the service. Four types of stainless are very common: ferritic stainless steel; austenitic stainless steel, martensitic stainless steel and precipitation hardening steel. Each stainless steel is composed of a different matrix (phase), and accordingly, each one offers different tensile strength, ductility, hardness, toughness, thermal expansion coefficient, high temperature, and corrosion resistance.

Table 6.1:. Effect of alloying element in steel on microstructure, mechanical properties, and respective metal strengthening mechanisms

Alloying element	Strengthening Mechanism	Microstructure	Mechanical properties

Tungsten (W)	Solid solution formation, hardenability, and compound formation	Martensite transformation, and WC formation	Increases strength, hardness at high-temperatures, wear resistance and hardness
Chromium (Cr)	Solid solution formation, hardenability, and compound formation	Martensite transformation, $Cr_{23}C_6$ formation	Increase hardness, high-temperature strength, wear resistance and corrosion resistance
Silicon (Si)	Grain refinement and hardenability	Transformation of single-phase austenitic to duplex microstructure containing ferrite and austenite.	Increase oxidation resistance and hardness
Molybdenum (Mo)	Solid solution formation, hardenability, and compound formation	Martensite transformation, Mo_2Cformation	Increase high-temperature strength, creep strength and hardness
Vanadium (V)	Solid solution formation, grain refinement, hardenability, and compound formation	Martensite transformation, VC/V_4C_3 formation	Increases high-temperature strength, creep strength, endurance limit and hardness
Nickel (Ni)	Solid solution formation, hardenability, and grain refinement	Martensite transformation, austenite stabiliser	Increase strength and toughness
Manganese (Mn)	Solid solution formation, compound formation, and hardenability	Martensite transformation, MnS formation	Increase hardness, toughness, abrasive resistance and ductility
Cobalt (Co)	Solid solution formation	Austenite stabiliser promotes the formation of carbides of other metals indirectly	Increase impact strength, and hot hardness
Titanium (Ti)	Solid solution formation, grain refinement, and compound formation	TiC formation	Increase inter-granular corrosion resistance and strength

Aluminium (Al)	Solid solution formation and grain refinement	Formation of aluminium nitrate	Increase oxidation resistance and strength

6.4.1 Plain Carbon Steel

These are the most common, popular, readily available, and low-cost steels. Mild steel, structural steel and a few types of tool steel are plain carbon steels. In these steels, only carbon concentration is controlled to achieve the desired microstructure in form of the relative amount of ferrite and pearlite which in turn helps to realize the mechanical properties namely hardness, tensile strength, ductility, and toughness. For example, the low carbon content in plain carbon steel results in a high proportion (e.g. 70-90%) of soft and ductile ferrite and a limited fraction (e.g. 10-30%) of hard and strong pearlite. Such a combination of ferrite and pearlite leads to relatively low hardness, low yield strength, high ductility and high toughness. On the other hand, the high carbon content in plain carbon steel produces a high proportion (e.g. 70-100%) of hard and strong pearlite and a limited amount (e.g. 0-70%) of soft and ductile ferrite. This type of ferrite and pearlite combination causes relatively high hardness, high yield strength, low ductility, and low toughness. Medium carbon steel is composed of moderate microstructure (e.g. 40-60% pearlite and ferrite as per carbon content) and mechanical properties accordingly. In light of service conditions (concerning external load), steel of a suitable combination of mechanical properties (hardness, yield strength, ductility, and toughness) is selected. For example, low-carbon steel is preferred in impact load conditions. In contrast, high-carbon steel is chosen for high resistance to plastic deformation and wear conditions. Table 6.2 shows the common plain carbon steels and their applications.

6.4.2 Alloy steel

Alloy steel is designed by adding different alloying elements (Cr, Mo, Al, V, W, Mn, Ni, etc.) to obtain the desired microstructure and mechanical properties in rolled or heat-treated conditions.

These alloying elements modify the microstructure and mechanical properties in different ways. The elements like aluminium, vanadium, and titanium refine the microstructure of alloy steel. Most of the alloying elements (except Co) including carbon reduce the critical cooling rate and increased the hardenability of steel which in turn allows the strengthening of alloy steel even by air hardening (austenite to martensite transformation). The presence of a few elements helps in the transformation of austenite to bainite transformation. A few elements (Mn, Si, C), on the other hand, simply form the solid solution to strengthen the steel, while other elements form the hard and stable precipitates, carbides, and intermetallic compounds to strengthen the alloy Steel. Since the different alloying elements act in different ways (solid solution, grain refinement, precipitation, strain, and transformation hardening singly or in combination) to strengthen the steel, relative effect of these alloy elements to strengthen the steel also varies significantly. Table 6.1 shows the effect of different alloying elements on the microstructure and mechanical properties of alloy steel.

6.4.3 Stainless steel

The matrix phase (ferrite, austenite, martensite, or mix of phases) of stainless steel is primarily determined by its chemical composition. All the alloying elements (including carbon and chromium) present in stainless steel are placed in two categories. One category belongs to all those elements (Cr, V, W, Mo, etc.) which tends to stabilise the ferrite called ferrite stabilisers and expressed in terms of Cr-equivalent, and another one includes all those elements, which stabilise the austenite called austenite stabilisers and expressed in terms of Ni-equivalent. The Cr and Ni equivalents of stainless steel are calculated using equations below where the element shows wt.% of that particular alloying element. Schaeffler diagram helps to estimate phase(s) present in stainless based on Ni and Cr equivalent (from the chemical composition) as shown in Fig. 1. A point is located on the Schaeffler diagram using Ni and Cr equivalent, which can indicate single phase A (Ferrite, Austenite, Martensite only), two phases B (Ferrite-Martensite, Ferrite-Austenite, Austenite-Martensite) and three (Ferrite-Austenite-Martensite)

Cr equivalent: $Cr\% + Mo\% + 1.5Si\% + 0.5Nb\%$Ni equivalent: $Ni\% + 30\ C\% + 0.5Mn$

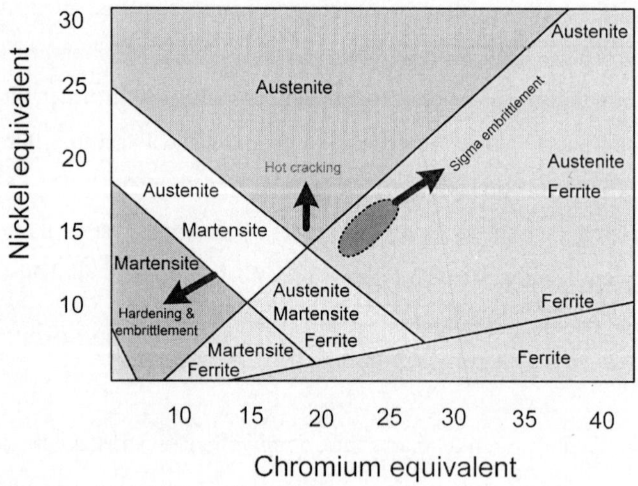

Figure: 6.1 Schaeffler diagram showing various matrix phases as per Cr and Ni equivalents

6.4.3.1 Ferritic stainless steel

The ferritic stainless steel (as the name suggests) comprises a matrix of ferrite apart from other phases. Ferritic stainless steel primarily consists of ferrite-stabilising elements like chromium, vanadium, and tungsten. The carbon content in FSS is kept below 0.05% C. The ferritic stainless steel offers moderate resistance to corrosion and high-temperature conditions. The ferritic stainless steel, however shows poor formability and weldability. Exposer of ferritic stainless steel to a typical thermal cycle during the fusion welding degrades the toughness and resistance to the corrosion of heat affected zone. The coarsening structure and hard, brittle notch-sensitive sigma phase formation in the heat-affected zone of high Cr ferritic stainless steel decreases the toughness and fatigue resistance. Further, chromium carbide formation in the heat-affected zone causes weld decay. The weld decay in HAZ of The ferritic stainless steel occurs at a little distance from the fusion boundary. The weld decay decreases the corrosion resistance and promotes pitting and galvanic corrosion in the halide environments like NaCl.

6.4.3.2 Martensitic stainless steel

The martensitic stainless steel (as the name suggests) has a martensite matrix apart from other phases. Martensitic stainless steel contains a higher concentration of carbon and a lower amount of Cr than ferritic stainless steel, besides other minor elements like chromium, molybdenum, vanadium, tungsten, etc. Martensitic stainless steel shows higher yield strength, hardness, and wear resistance than ferritic stainless steel while offering moderate resistance to corrosion and high temperature. Martensitic stainless steel is significantly used for making surgical tools and cutlery due to good corrosion resistance and hardness. Similar to ferritic stainless steel, the weld thermal cycle during the fusion welding decreases the toughness and resistance to the corrosion of heat affected zone of martensitic stainless steel due to grain coarsening and the formation of hard and brittle martensite. Further, chromium carbide formation in heat-affected zones causes weld decay in martensitic stainless steel, decreasing corrosion resistance and promoting galvanic corrosion in halide environments.

6.4.3.3 Austenitic stainless steel

The austenitic stainless steel (as the name suggests) is composed of an austenite matrix apart from other phases. Apart from Cr, austenitic stainless steel (ASS) contains a significant amount of austenite-stabilising elements like nickel, manganese etc. The carbon content in ASS is kept below 0.05% C, and it is preferred to have carbon even less than 0.02 % to avoid weld decay and hair/knife line cracking.

The austenitic stainless steel offers very good corrosion resistance and can withstand high temperatures (up to about 600 °C). The austenitic stainless steel shows very high toughness (150 to 220 J) and ductility (70-80 %) resulting in good formability. The weldability of austenitic stainless steel using gas tungsten arc welding (GTAW) and gas metal arc welding (GMAW) processes is good. Autogenous welding and fusion welding using inappropriate filler/electrode cause solidification cracking at the weld centreline of the ASS weld joint. However, heat affected zone of austenitic stainless steel suffers from weld decay, hair/knife line cracking, and embrittlement due to the formation of hard, brittle, and notch-sensitive sigma phase. Moreover, a high thermal expansion coefficient (almost twice of carbon steel) of austenitic stainless steel promotes residual and distortion tendency. These issues suggest that a suitable welding procedure has to be developed to make a weld joint free from discontinuities. The coarsening structure and hard, brittle notch-sensitive sigma phase formation in the heat-affected zone of high Cr (>20 wt.%) austenitic stainless steel decreases the toughness and fatigue resistance. The formation of chromium carbide in the heat-affected zone leading to weld decay is one of the most common issues in fusion welding of weld decay. The weld decay in HAZ of austenitic stainless steel is observed away from the fusion boundary. The weld decay decreases the corrosion resistance and promotes galvanic corrosion, especially in halide environments. The weld decay of all types of stainless steels can be addressed using various approaches like heat treatment, low carbon stainless steel (<0.02 %C), and stabilising the stainless steel (by alloying stronger carbide formers (like Nb, Ti) than chromium. Heat treatment involves heating the weld joint at 1000-1100 °C for one hour, followed by rapid cooling. High-temperature exposure causes the dissolution of chromium carbide (precipitated at the grain boundaries of the austenite) in the heat affected zone. In stabilised stainless steel, the presence of Nb and Ti being stronger carbide formers forms their carbide and avoids chromium carbide formation, leading to improved resistance to weld decay. Since weld decay is linked with chromium carbide formation, therefore, a reduction in carbon content in stainless steel (in the form of low carbon stainless steel) improves the corrosion resistance. For example, low carbon austenitic stainless steels (AISI 304 L, AISI 316 L) having lesser carbon (<0.02%) than the normal austenitic stainless steel (AISI 304, AISI 316 with 0.05% C) offers better corrosion resistance.

Table 6.2:. AISI steel designation, composition, and applications

Category	Steel	Designations	Composition[a](wt.%)	Applications
Plain C steels	Low carbon steels	AISI 1010	0.10 C, 0.45 Mn	Automobile panels, nails, and wire
		AISI 1020	0.20 C, 0.45 Mn	Pipe; structural and sheet metal
		A36	0.29 C, 1.00 Mn, 0.20 Cu (Min.)	Structural (bridges and buildings)
		A516 Grade 70	0.31 C, 1.00 Mn, 0.25 Si	Low-temperature pressure vessels
	Medium carbon steels	AISI 1030	0.30 C, 0.75 Mn	Machinery parts, clutches, brakes
		AISI 1040	0.40 C, 0.75 Mn	Crankshafts, bolts
	High carbon steels	AISI 1080	0.80 C, 0.75 Mn	Chisels, hammers
		AISI 1095	0.95 C, 0.75 Mn	Knives, hacksaw blades
Alloy steels	Low alloy steels	AISI 4063	0.63 C, 0.25 Mo	Springs, hand tools
		AISI 4340	0.40 C, 1.80 Ni, 0.80 Cr, 0.25 Mo	Bushings, aircraft tubing
		A440	0.28 C, 1.35 Mn, 0.30 Si (max.), 0.20 Cu (min.)	Structures that are bolted or riveted

	Micro-alloyed steels(HSLA)	A633 Grade E	0.22 C, 1.35 Mn, 0.30 Si, 0.08V, 0.02 N, 0.03 Nb	Structures used at low ambient temperatures
		A656 Grade 1	0.18 C, 1.60 Mn, 0.60 Si, 0.1 V, 0.20 Al, 0.015 N	Truck frames and railways cars
Stainless steels	Ferritic stainless steels	AISI 409	0.08 C, 9.0 Cr, 1.0 Mn, 0.50 Ni, 0.75Ti	Automotive exhaust components, tanks, agricultural sprays
		AISI 430	16% Cr, 0.12% C	
		AISI 446	0.20 C, 25 Cr, 1.5 Mn	Valves (high temperature), glass molds, combustion chambers
	Martensitic stainless steels	AISI 410	0.15 C, 12.5 Cr, 1.0Mn	Rifle barrels, cutlery, jet engine parts
		AISI 416	0.12 C, 11.5 Cr, 1.25 Mn	Valves, motor shafts, gears, bolts, nuts
		AISI 440A	0.70 C, 17Cr, 0.75Mo, 1.0Mn	Cutlery, bearings, surgical tools
	Austenitic stainless steels	AISI 304	0.08 C, 19Cr, 9 Ni, 2.0Mn	Chemical and food processing equipment, cryogenic vessels
		AISI 316L	0.03 C, 17Cr, 12 Ni, 2.5 Mo,2.0Mn	Welding construction
	PH-stainless steels	17-74H	0.09 C, 17Cr, 7 Ni, 1.0 Al, 1.0Mn	Springs, knives, pressure vessels

6.5 Tool and Die Steel

Steel for making forming and cutting tools and dies for desired performance in terms of reliability and life should have a good combination of hardness, wear resistance, toughness, dimensional stability, and ease of manufacturing. In general, increasing the hardenability of the steel increases the maximum attainable hardness, wear resistance, and dimensional stability. Depending upon the hardening approach and application, tool steel are classified as water hardening (plain carbon steel), shock resisting (medium and low alloy steel), cold work (oil hardening, high carbon-high chromium, medium alloy steel), hot work (chromium, molybdenum, tungsten Steel), high-speed steel (tungsten, and molybdenum high-speed steels), and mold (low carbon steel).

6.5.1 Water hardening steels

These are plain carbon steels having carbon in the range from 0.5 to 1.4% with a small addition of alloying elements like chromium and vanadium to improve the toughness and wear resistance. Water-hardening steel performs well as a cutting tool under moderate temperature/heat generation and forces during operations.

6.5.2 Shock resisting steel

Compared to water-hardening steel, shock-resistant steel contains lesser carbon (0.4 to 0.6%) and a higher concentration of alloying elements like silicon, chromium, tungsten, and molybdenum. Each element plays a specific role. Silicon forms a solid solution to increase the strength of ferrite, chromium, and molybdenum, increasing the hardenability, while tungsten increases the hot hardness of the steel. The presence of these alloying elements increases the hardenability; therefore, this steel allows air-hardening. The maximum attainable hardness of this steel is about 60 HRC. Punches, chisels, forming tools, and blades are made of shock-resisting steel.

6.5.3 Cold work steel

These are primarily air-hardening low alloy steel with very low carbon content (~1%) and 5% Cr, 2% Mn, and 1% Mo. A combination of carbon and high chromium in this steel produces good hardness and wear resistance. These Steels offer good wear resistance and resistance to decarburisation, while toughness and hot hardness are moderate. These are mainly used for making the punches and dies for blanking, forming, trimming, etc.

6.5.4 Hot work steel

The tools and components used in hot working (dies, rollers) and the foundry industry (die casting mild) exposed to the high temperature during the service must have enough hot hardness to resist thermal softening and ensure good dimensional stability. Adding alloying elements like chromium molybdenum tungsten up to 5% in steel increases the hot hardness. Accordingly, there are three categories of hot working steels i.e., Chromium type, tungsten type, and molybdenum type. The chromium-type hot working steel contains 3.5% chromium with a small amount of vanadium, tungsten, and molybdenum. The hardness of this type of steel is found in the range of 42-55 HRC, along with a good combination of toughness. The tungsten type of hot working steel contains 9 to 18% of tungsten and 3 to 12% of chromium. The toughness of the tungsten type steel is better than the chromium type hot working steel. The characteristics of molybdenum-type hot-working steel are similar to that of tungsten-type hot-working steel. However, molybdenum-type steels are cheaper than tungsten-type of hot-working steels.

6.5.5 High-speed Steel

These steels are highly alloy steels containing tungsten/molybdenum along with other elements like chromium and vanadium. Carbon content varies point 0.7 to 1.2%. 18-4-1 is one of the most common types of high-speed steel used in industries where 18 is the tungsten %, 4 is the chromium %, and 1 is the vanadium %. These steels are mostly used for making burnishing tools, blanking tools, extrusion dies, tool bits, drills, reamers, broaches, taps, milling cutters, hobs, and woodworking tools. Sometimes even cobalt is added to increase the hot hardness of high-speed steel. There are two broad categories in high-speed steel (based on the composition) molybdenum type (M Type) and tungsten type (W type). M-type high-speed steel is cheaper than the W-type high-speed steel, while the performance of both types of high-speed steel is almost the same.

6.5.6 Mould Steel

These are low Carbon steel with a small amount of Cr and Ni. These steels are case hardened by carburising to increase the surface hardness from 55 to 64 HRC. The hot hardness of these steels is relatively poor. These are mostly used for making the dies for metal casting and making moulds for injection and compression moulding of the plastics.

6.6 Maraging steel

Maraging steel is a low-carbon martensitic and precipitate-hardened steel. It contains substitutional alloying elements. The word, maraging is derived from the combination of martensite and age hardening. The steel is subjected to artificial aging treatment to fine precipitates, which act as obstacles for the movement of dislocations. The selection of aging conditions (temperature and time) both are important for peak age hardening. Over-aging and under-aging both decrease the strength and hardness of maraging steel. Over-aging (due to high aging temperature exposure for a long time) coarsen the precipitates, while under-aging (due to low temperature and short time for aging) results in insufficient precipitate formation.

The low carbon martensite with hardening precipitates in maraging steel results in high strength, ductility, impact strength, toughness, and fracture toughness. It is very unique steel that offers very high tensile strength and high fracture toughness while other steels usually offer high strength but low fracture toughness or vice versa. The high strength-to-weight ratio makes this steel suitable for aerospace, automobile, and military application.

The maraging steel is an alloy of iron and elements like nickel (17-19%), cobalt (8-12%), molybdenum (3-5%), titanium (0.2-1.8%), aluminium (0.1-0.15%), and very less percentage of carbon (0.03%). The low carbon content avoids the formation of titanium carbide, which is known to decrease the toughness, ductility, and impact strength of this steel. Maraging steel is more expensive than other steels due to the presence of cobalt as an alloying element. Maraging steel is known by its nominal yield strength ranging from 150 to 350 ksi (1030 to 2040 MPa) which may be further higher up to 500 ksi (3450 MPa) due to the addition of Cobalt. Common maraging steels with respective composition, aging temperature, time and mechanical properties are tabulated in Table 6.3 and 6.4.

Table 6.3: Composition and treatment of common maraging steels

Grade	Composition (wt. %)					Aging temperature in °C and (time, hr)
	Ni	Mo	Co	Ti	Al	
18Ni(200)	18	3.3	8.5	0.2	0.1	480 (3)
18Ni(250)	18	5.0	8.5	0.4	0.1	480 (3)
18Ni(300)	18	5.0	9.0	0.7	0.1	480 (3)
18Ni(350)	18	4.2	12.5	1.6	0.1	480 (12)
18Ni(Cast)	17	4.6	10.0	0.3	0.1	-
12-5-3(180)	12	3.0	0	0.2	0.1	480 (5)

Table 6.4: Mechanical properties of common maraging steels

Grade	Tensile strength MPa (ksi)	Yield strength MPa (ksi)	Elongation (%)	Fracture toughness (MPa $m^{1/2}$)	Toughness (J)
18Ni(200)	1500 (218)	1400 (203)	10	140	60
18Ni(250)	1800 (260)	1700 (274)	8	100	35
18Ni(300)	2050 (297)	2000 (290)	7	70	25
18Ni(350)	2450 (355)	2400 (348)	6	40	16
18Ni(Cast)	1750 (255)	1650 (240)	8	-	-

12-5-3 (180)	1895 (275)	1825 (265)	11.5	-	-

6.6.1 Embrittlement of maraging steel

In the maraging steel is designed to have low carbon content (<0.03%). Higher carbon content (>0.02%) doesn't help to avoid embrittlement because at high temperature, carbon atoms starts diffusing into austenite and accumulating at the grain boundaries during slow cooling or holding of the steel ,resulting into embrittlement.

Overheating of high carbon (more than 0.03 %) maraging steel to high temperature (>1180 °C) followed by slow cooling (through critical temperature range 750 °C-1050°C) causes interaction of Ti with carbon and nitrogen which in turn forms titanium carbide/titanium carbo-nitride. These carbides and nitrides precipitate at the grain boundaries of austenite causing embrittlement of maraging steel. Such embrittlement of steel can be neutralized by heat treatment involving the heating of the steel to 1150°C followed by rapid cooling down to 750°C to avoid precipitation of carbide and carbo-nitrides at grain boundaries of austenite causing embrittlement.

6.6.2 Physical metallurgy of Maraging Steel

Maraging steel is formed after heating the steel up to the austenitic phase (~850°C), followed by air-cooling to convert austenite into martensite. Air hardening of maraging is due to high hardenability caused a high concentration of alloying elements. Artificial aging of air-hardened martensite at about 480-500°C for few hours results in hardening precipitates like Ni_3Mo, Ni_3Ti, Ni_3Al, and Fe_2Mo in soft martensite matrix. Low carbon content (<0.02%) in this steel during aging produces very negligible carbide formation.

Cobalt in the maraging steel is a major alloying element, which accelerates precipitation of well-distributed precipitates (Fe_2Mo, Ni_3Mo) in the matrix of martensite, and reduces the aging time for peak hardening.

6.6.3 Applications of maraging steel

Toughness, ductility maraging steel due to the presence of soft martensite shows good machinability. Therefore, it suits applications involving high stress coupled with impact loading. Following are a few common applications of the maraging steel.

Structural component: Parts of the landing gear of aircraft and spacecraft, the undercarriage of aeroplanes, rocket/missile motor case, anchor rails, arresting hooks, universal flexural, Nuclear plants (rotors, shafts)

Machine tool component: Load cells, piston, mandrel for cold tube drawing, gear for machine tools, extrusion: ram, die container, press, hydraulic press, carbide die holder, die used for casting and forging process.

Artillery component: Cannon recoil springs, light weight portable springs,

Automobile component: Shaft, gears, rod, and pistons mainly for racing cars.

6.7 Cast Irons

Cast iron is an alloy of iron, carbon (>2.0%), and silicon (1-3%) where phosphorus and sulphur are present as impurities. Silicon in cast iron promotes graphitization (carbon in free form). Adding other alloying elements (Cr, Mo, Ni Cu etc.) primarily helps achieve specific mechanical and tribological properties. The compressive strength, hardness wear resistance of cast iron are significantly higher than steel. In general, the toughness and ductility of cast iron are significantly lower than steel. The mechanical properties of cast irons are primarily governed by the microstructure, including the form in which carbon is present and the phase of the matrix. Carbon can be present in a combined form like Iron carbide or a free form like graphite. Similarly, the cast iron matrix can be ferritic, pearlitic, austenitic, and martensitic, depending upon the composition and cooling conditions imposed during the solidification and thermal treatment. The strength and ductility of the cast iron are significantly influenced by the relative amount of free carbon and carbon in combined form besides its morphology and size. Increasing carbon in combined form (iron carbide, alloy carbide) and reducing the grain size of phases/compounds in spherical morphology increase strength and ductility. Fine micro-constituents of spherical morphology (as compared to high aspect ratio phases like a needle, and flakes) result in higher toughness and ductility. The heat treatment (like normalizing, annealing, hardening, and tempering) of cast iron can also help to achieve the desired microstructure and mechanical properties if not realized in the as-cast condition. The classification of cast iron (gray, malleable, nodular and white cast iron) is based on the appearance of graphite under the optical microscope in polished conditions.

6.7.1 Gray cast iron

It contains carbon in free form as graphite flakes (Fig. 6.2). Gray cast iron is named so because the appearance of the fracture surface of this cast iron is gray. Mechanical properties and microstructure of the grey cast iron depend on chemical composition and cooling rate experienced by the molten metal during the solidification. The addition of alloying elements like copper, chromium, and molybdenum helps to achieve the desired microstructure of the matrix and graphite. The hardening of the cast irons can be achieved through quenching treatment like steel followed by tempering to induce the desired toughness and ductility but certainly at the cost of hardness. The graphite flakes offer negligible strengthening to cast iron, while high-stress concentration at the flake-matrix interface promotes crack nucleation and reduces ductility.

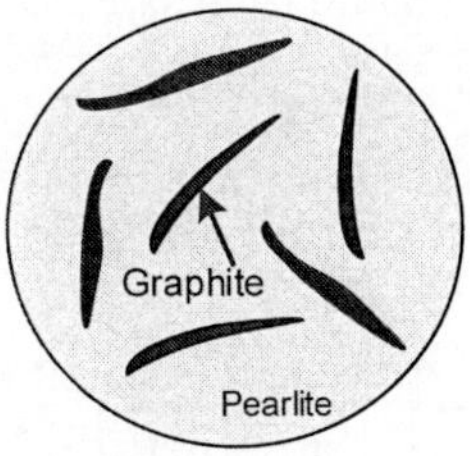

Figure 6.2: Schematic diagram of an optical micrograph of grey cast iron showing flakes and pearlite matrix

6.7.2 Nodular cast iron

Nodular cast iron is similar to gray cast iron with regard to the composition (C, Si) however, this cast iron defers in terms of the morphology of the graphite. The nodular cast iron consists of compact nodules of graphite in a spherical morphology (Fig. 6.3). The spherical mythology of graphite reduces the stress concentration at the graphite-matrix interface, which in turn results in higher strength, toughness, and ductility than the gray cast iron. Speckle morphology of the graphite is achieved by admin magnesium or cerium in the molten metal during the casting (<0.02% S). The magnesium (~0.035%) must be added carefully in the molten metal to avoid evaporation. Ferro-magnesium master alloys can be used for this purpose.

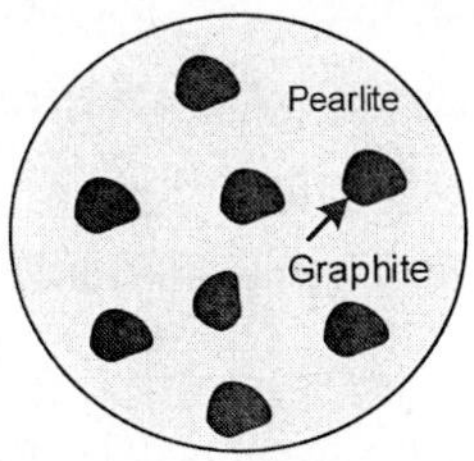

Figure 6.3: Schematic diagram of an optical micrograph of nodular cast iron showing graphite nodules and pearlite matrix

6.7.3 White cast iron

White cast iron contains carbon in a combined form like iron carbide and other alloy carbides as per composition (Mo, Cr), as shown in Fig. 6.4. The development of white cast iron needs proper control of the chemical composition (regarding the carbon and silicon content), alloying elements (Ni, CU, Cr, Mo), and cooling rate during the solidification. Low carbon and silicon content in the presence of suitable alloying elements result in white cast iron, which is composed of alloying carbides and a hard martensitic matrix. This is the hardest type of cast iron, with almost no ductility and toughness. The fracture surface of this cast iron gives a white crystalline appearance therefore; it is called white cast iron. White cast iron offers excellent abrasive wear resistance. White cast iron shows a high cracking tendency during the welding due to the minimal toughness and ductility.

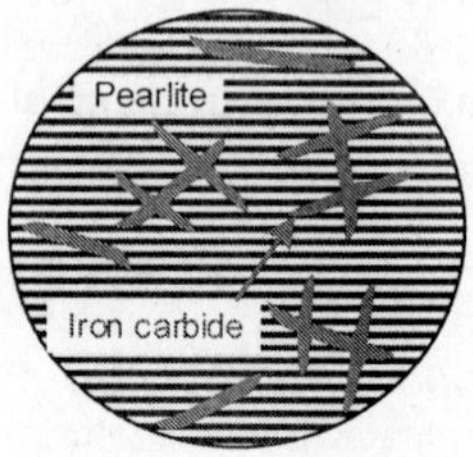

Figure 6.4: Schematic diagram of an optical micrograph of white cast iron showing iron/alloy carbides and pearlite matrix

6.7.4 Malleable cast iron

The malleable cast iron has free carbon in the form of regular shape graphite to achieve the desired ductility and impart malleability at high temperatures (Fig. 6.5). The malleable cast iron is obtained through the heat treatment of white cast iron. Heating the white cast iron at the eutectoid temperature for a long time causes the decomposition of iron and alloy part carbides which eventually precipitates as regular shape graphite particles in the matrix of pearlite or martensite as per the composition of white cast iron. The strength and ductility of the malleable cast iron are determined by the microstructure of the matrix and the morphology of the irregular graphite particles. To restore the properties of the white cast iron (from malleable cast iron), a hardening heat treatment may be needed.

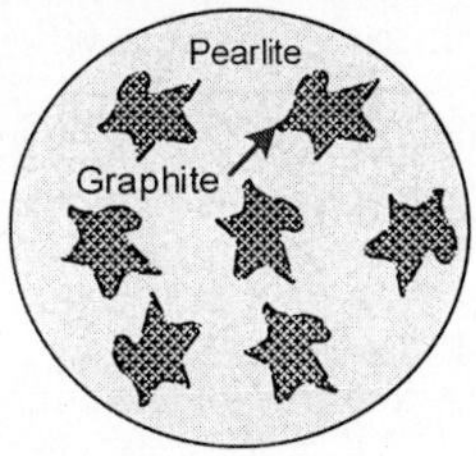

Figure 6.5: Schematic diagram of an optical micrograph of nodular cast iron graphite nodules and pearlite matrix

6.7.5 *Physical metallurgy of cast irons*

Cast iron is an alloy of the iron-carbon system which melts at the lowest temperature. The low melting temperature of cast iron and other attractive characteristics like low thermal expansion coefficient makes it easy to cast. Therefore, it is called cast iron. High carbon content in cast iron leads to increased hardness and compressive strength but low toughness and poor ductility. However, the mechanical properties of cast iron significantly depend on two aspects a) the form in which carbon is present in cast iron and b) the morphology of carbon/its compound. Carbon in cast iron can be present in free or combined form.

Carbon in free form in cast iron appears as graphite which is soft and acts solid lubricant. The free form of carbon (graphite) morphology can be nodules, flakes and irregular shapes in nodular, grey and malleable cast iron, respectively. Carbon can also be present in the combined form of "iron carbon or alloy carbide". The most common form of cast iron in the combined form of carbon (iron carbide) is white cast iron. The carbon in combined form results in very high hardness and abrasive wear resistance but almost negligible toughness and ductility.

Alloying elements and cooling rate experienced by cast iron during the solidification significantly affect the type of cast iron. The presence of alloying elements like W, V, Cr increasing hardenability at high cooling in variably produces white or alloy cast iron. Few alloying elements like Si promote the graphitization to produce grey cast iron. An alloying element like Mg in molten cast iron modifies the morphology of graphite flake to nodules leading to the nodular / spheroidal cast iron. Heat treatment of grey cat iron produces the irregular shape of graphite to make malleable cast iron.

6.8 Copper and copper alloys

Copper and its alloys are one of the most important engineering materials for the fabrication of electrical components, fluid handling systems, and heat exchangers due to attractive physical properties like high electrical and thermal conductivity coupled with excellent corrosion resistance against fresh and salty water, and many chemical environments. Copper in its pure form consists of a face-centred cubic structure, which provides good ductility, formability, and malleability. Copper (8.94 g/cc) is heavier than iron (7.8 g/cc) and aluminium (2.84 g/cc), and the electrical conductivity of copper is about 1.5 times higher than aluminium.

The electrical conductivity of copper is considered a standard (100%) among engineering materials. The electrical conductivity of other materials is rated with respect to that of copper (as per ICAS International copper annealed standard). Copper is frequently alloyed with common elements like aluminium, nickel, silicon, tin, and zinc to improve its mechanical properties, machinability, corrosion resistance, and response to heat treatment. Based on composition, copper and its alloys are grouped into nine categories, namely

- Commercial Cu (Min. 99.3 % Cu)
- High Cu alloy- 5% alloying element remaining copper
- Cu-Zn alloy (brass)
- Cu-Sn (Phosphor bronze)
- Cu-Al (Aluminium bronze)
- Cu-Si (Silicon bronze)
- Cu-Ni alloy (Nickel bronze)
- Cu-Ni-Ag alloy
- Special alloys

The addition of alloying in copper affects the microstructure and mechanical properties, melting temperature and solidification temperature range, thermal and electrical conductivity, thermal expansion behaviour, corrosion resistance, and work hardening tendency, which in turn determine the ease of manufacturing, mechanical performance, and corrosion behaviour of copper alloys (Table 6.5).

Any kind of alloying in pure copper decreases electrical and thermal conductivity and ductility. The addition of alloying elements like iron, silicon, arsenic, tin, and antimony increases the corrosion resistance of copper alloys. Alloying copper with lead, selenium, and tellurium improves machinability. During the casting of copper alloys, adding boron, phosphorus, silicon and lithium helps refine the grain structure. The addition of cadmium and silicon increases the hot hardness of copper alloys. The precipitation hardening can be realised in copper alloyed with Cd, Co, Zr, Cr, and Be to increase the strength. Single-phase copper alloys (solid solutions) allow manufacturing through cold working easily as two-phase copper alloys work hardens rapidly. An increase in the percentage of hardening precipitates increases the yield strength but at the cost of ductility of copper alloys. Similarly, strength and ductility of copper alloy are significantly affected by temperature due to thermal softening caused recovery and recrystallization (Fig. 6.6).

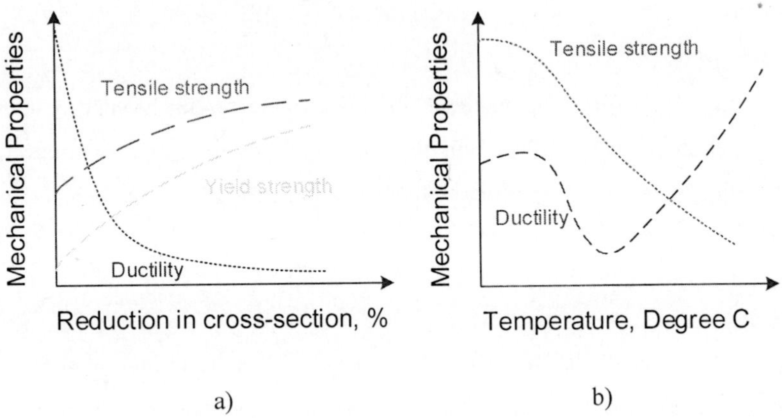

Figure 6.6: Schematic diagram showing variation in mechanical properties of copper alloy as a function of a) reduction in cross section due to plastic deformation and b) temperature

6.8.1 Effect of alloying elements

Aluminium

Aluminium affects the microstructure and mechanical properties. The addition of aluminium up to 8% forms single-phase solid solution with copper while further the addition of the Al 9-15% results in the two-phase Cu-Al alloy. system. In general, increasing aluminium content increases strength and hardness but at the cost of ductility. Aluminium in copper forms a refractory aluminium oxide layer which interferes in the melting of the Cu-Al alloy.

Arsenic

Alloying of copper with arsenic reduces dezincification (corrosion) of Cu-Zn alloy.

Beryllium

The high solubility difference of beryllium in copper at room (0.3%) and high temperature (2%) makes Cu-Be alloy precipitation hardenable. The addition of beryllium in copper decreases the thermal and electrical conductivity and melting point, making the casting and welding of Cu-Be alloy little easier.

Boron

The addition of boron in copper acts deoxidiser which improves the weldability of copper alloys.

Cadmium

A minor edition (up to 1.25%) of the cadmium in copper increases the strength without compromising the electrical conductivity and therefore suits for making electrodes for resistance spot welding. However, the evaporation of cadmium during fusion welding affects the weld composition and mechanical properties.

Chromium

The difference in solid-state solubility of Cr in Cu at room (0.05%) and high temperature (0.55%) makes Cu-Cr alloy precipitation hardenable leading to good strength without significantly decreasing the electrical conductivity (Fig. 6.7). The addition of Cr in copper forms a refractory oxide layer (like Al, Be) which must be taken care of during the fusion welding.

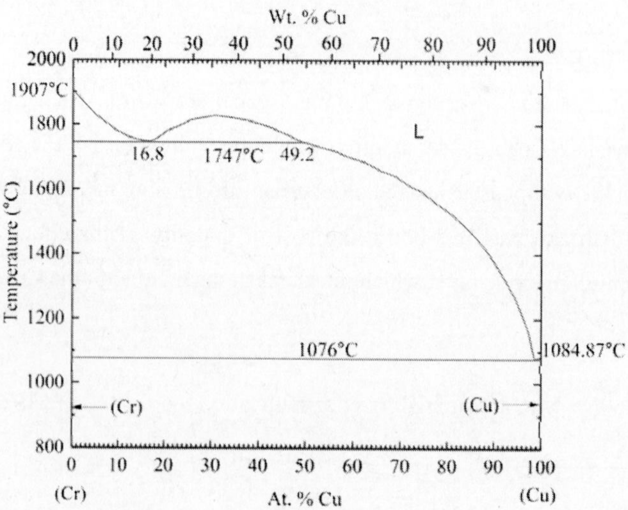

Figure 6.7 Binary phase diagram of Cu-Cr alloy

Iron

The addition of iron in Cu alloys (such as aluminium bronze, manganese bronze and copper Nickel alloys) increases strength, erosion, and corrosion resistance due to grain refinement, solid solution strengthening and precipitation hardening. Solid-state solubility of iron in copper is higher (3%) at elevated temperatures than the room temperature (0.1%).

Manganese

The addition of Mn in deoxidised copper, manganese bronze, and silicon bronze up to 3.0% forms a solid solution increasing the ease of manufacturing by the hot working.

Nickel

The addition of nickel in copper forms solid solutions to increase the strength and hardness of the Cu-Ni alloys.

Phosphorous

The addition of phosphorus in copper increases strength and acts deoxidiser. Solid-state solubility of iron in copper is higher (1.7%) at elevated temperatures than the room temperature (0.4%).

Silicon

The effect of silicon in Cu is similar to that of aluminium. Si acts deoxidiser and solid solution hardener. Solid-state solubility of Si in copper is higher (5.3%) at elevated temperatures than the room temperature (3.6%). Increasing Si content generally increases Cu alloy's strength, ductility, and malleability. Si in copper also forms a refractory oxide layer which interferes with the melting of the Cu-Si alloy.

Zinc

Zinc is one of the most important elements for copper alloys. Zinc has higher solid solubility at room temperature (37%) than at elevated temperature (32.5%). However, zinc is present in different copper alloys (aluminium bronze, copper-nickel alloy) and evaporates in the molten state, which in turn increases the porosity and cracking tendency.

Tin

Addition of tin in the copper results in various alloys. The solid solubility of tin in copper is higher (13.5%) at elevated temperatures than the room temperature (1%) as shown in Fig. 6.8. However, rapid cooling can result in a single phase even with 2% tin-copper alloys.

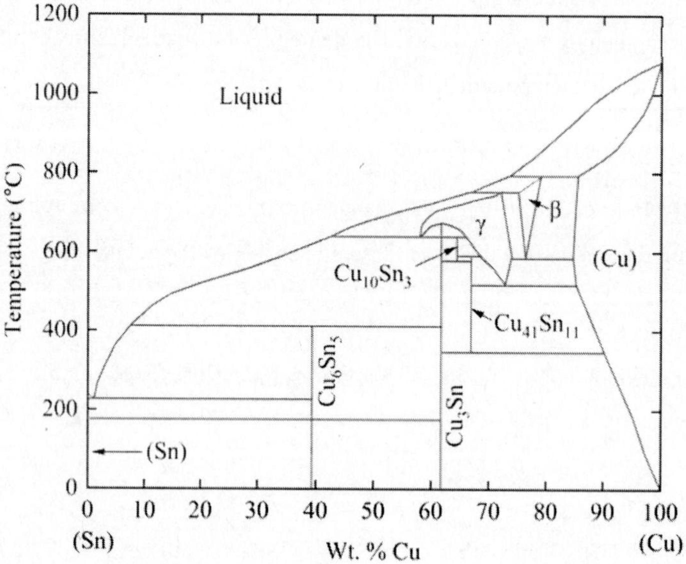

Figure 6.8 Binary phase diagram of Cu-Sn alloy

6.8.2 Alloying and solidification temperature

Solidification temperature range plays a crucial role in developing the fusion weld joint and castings free from hot / solidification cracks. A high solidification temperature range (> 50 Degree C) generally promotes solidification cracking in fusion weld joints along the weld centreline and hot tears in castings. The composition of the copper alloys determines the solidification temperature range.

Based on the solidification temperature range, the copper alloys are grouped into three categories, i.e. Group I, II, and III. The copper alloys having a narrow certification temperature range (50 Degrees C) are placed in group I. A few common copper alloys of this group are aluminium bronze, Nickel-bronze, manganese bronze, chromium-copper alloys, and pure copper. The copper alloys (Be-Cu, Silicon bronze, copper-nickel alloy) having moderate solidification temperature range from 50 to 110 degrees C are segregated in group II. Group III copper alloys (leaded-brass, Tin bronze, lead-tin bronze) have the widest certification temperature range (110 to 170 degrees C).

Table 6.5: Physical properties of common copper alloys

Alloy	Melting range, °C	Coeff. of thermal exp., μm/per °C (20° to 300°C)	Thermal conductivity at 20 °C, W/(m·K)	Electrical conductivity, percent, % IACS
Oxygen free copper	1066-1080	17.6	370	101
Beryllium copper	866-982	17.8	107-130	22
Commercial bronze, 90%	1021-1043	18.4	188	44
Red brass, 85%	988-1027	18.7	159	37
Phosphor bronze, 10%D	843-999	18.4	50	11
Aluminium bronze, D	1041-1046	16.2	67	14
High silicon bronze, A	971-1027	18.0	36	7
Manganese bronze, A	866-888	21.2	105	24
Copper-nickel, 30%	1171-1238	16.2	29	4.6
Nickel silver, 65-15	1071-1110	16.2	33	6

6.8.3 Copper alloys

6.8.3.1 Oxygen-free copper

The oxygen-free copper has a maximum of 10 PPM oxygen and is produced in a reducing environment to avoid contamination from oxygen in the atmosphere. The mechanical properties of oxygen-free copper are similar to that of oxygen-bearing copper. The ductility of oxygen-free copper is extremely good. Oxygen-free copper may be alloyed with silver to enhance strength without compromising electrical and thermal conductivity. The addition of silver in oxygen-free copper increases the resistance to thermal softening and creep resistance.

6.8.3.2 Oxygen bearing copper

There are two types of oxygen-bearing copper named based on the manufacturing approach a) electrolytic tough pitch copper, and b) fire-refined copper. Electrolytic tough-pitch copper contains lesser impurities (Bi, Pb, As) than fire-refined copper. The oxygen-bearing copper shows a greater tendency to form porosity in fusion weld joints due to the higher oxygen content. Residual oxygen is almost the same in both types of oxygen-bearing copper. High oxygen content in oxygen-bearing copper forms copper cuprous oxide eutectic. In the presence of hydrogen, cuprous oxide causes embrittlement of oxygen-bearing copper due to the segregation of the steam (formed by the interaction of hydrogen and copper oxide) at the grain boundaries. Heating of the oxygen-bearing copper during the subsequent manufacturing (welding, heat treatment, hot working) promotes the segregation of copper oxide at the grain boundaries which in turn degrades the strength and ductility.

6.8.3.3 Free machining copper

The addition of lead, tellurium, and selenium in copper improves machinability. These elements have limited solid-state solubility in the copper therefore, the elements are found in the form of fine discrete particles and stringers in the copper matrix, which help in producing the fragmented chips during the machining. Additionally, these elements act as inclusions to reduce strength and ductility. Heating of free machining steels during the welding and heat treatment increases the hot shortness and cracking tendency.

6.8.3.4 Precipitation hardening copper alloys

The addition of beryllium chromium and zirconium in copper makes them precipitation hardenable to increase the strength following T6 heat treatment comprising solutionizing, quenching, and artificial aging. Aging results in fine precipitates which help in increasing the hardness and strength. Heating of precipitation hardening copper alloys during welding causes over-aging, degrading the mechanical properties. Precipitation hardenable copper alloys are cold worked in solutionized condition.

6.8.3.5 Cu-Zn alloy

Copper-zinc alloys are also known as brasses where zinc is a major alloying element (up to 50%). Minor addition of the elements like manganese, titanium, iron, silicon, nickel lead aluminium in brass improves the mechanical properties and corrosion resistance. The brasses are grouped into three categories, namely low zinc brasses (<20%), high zinc brasses (>20%), and leaded brass. On the other hand, considering the microstructure, there can be alpha, beta, and alpha-beta brass as per the zinc content (Fig. 6.9). Thus, zinc content significantly determines the microstructure, mechanical properties, appearance, and physical properties (melting point, density, thermal and electrical conductivity). In general, increasing zinc content in brasses increases strength, hardness, ductility, and coefficient of thermal expansion, while density, melting temperature, and thermal and electrical conductivity are compromised. Adding zinc to copper changes the colour from red to golden and yellow. The brass up to 36% of zinc forms a single alpha phase in the microstructure, which is formed sensitive to heart cracking. Zinc content greater than 36% in brass results in Alpha Beta face, improving the hot working characteristics.

Brass in general shows high thermal and electrical conductivity, strength, formability and corrosion resistance. These characteristics make it suitable for heat exchanger and air-conditioning, automobile, marine and aerospace industries. The high corrosion resistance makes brass attractive for pipelines, valves, and fittings to transport potable water, industrial waste (saline solutions, alkaline solutions and organic chemicals) and gases.

α- Brass

A substitution solid solution of zinc in copper makes the alpha phase with FCC crystal structure. Alpha brasses contain less than 37-39% zinc. With increasing content, the colour of the alpha brasses changes from red to yellow, coupled with a decrease in ductility, elongation and corrosion resistance. This kind of brass is more malleable and ductile than other brasses. Therefore, it can be cold worked, which suits forming and solid state joining. The most popular alpha brass is composed of 70%Cu- 30%Zn termed as cartridge brass.

β-Brass

Beta Brass contains 45-50% zinc. The crystal structure of beta phase is BCC. Beta Brass is harder and stronger than alpha and alpha-beta brasses. Components of beta brass are made using hot working or casting. It is a somewhat less popular type of brass.

α-β Brass

Alpha-beta brasses have heterogeneous crystal structures, i.e. mix of alpha and beta phases, therefore, these are called duplex brasses. These brasses contain 37-45% Zn, however, proportions of the alpha phase and beta phases depends upon the zinc content. The addition of elements like Al, Si and Sn promotes the beta phase. These brasses respond to the heat treatment due to the presence of the β phase. Alpha-beta brass is harder and stronger than alpha brass. Due to the ease of forming at elevated temperatures, Alpha-beta brasses are known as hot-working brasses.

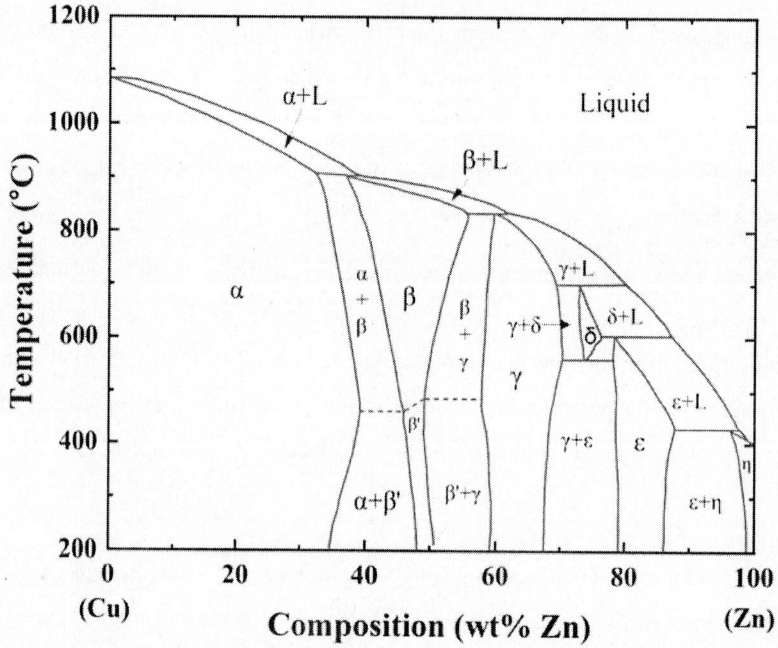

Figure. 6.9: Binary phase diagram of Cu-Zn alloy

6.8.3.6 Cu-Ni alloys

The nickel content in copper-nickel alloys ranges from 5 to 45%, along with the addition of minor alloying elements like Iron, magnesium and Zinc as shown in Fig. 6.10. Increasing Nickel content in these copper alloys increases strength and hardness; moreover, these alloys are known to be ductile and tough. However, the thermal and electrical conductivity of the copper-nickel alloy is low. The sulphur, lead, and phosphorus presence in copper-nickel alloys increases embrittlement. The addition of silicon in copper Nickel alloy increases the fluidity of the molten metal and the strength of the alloy.

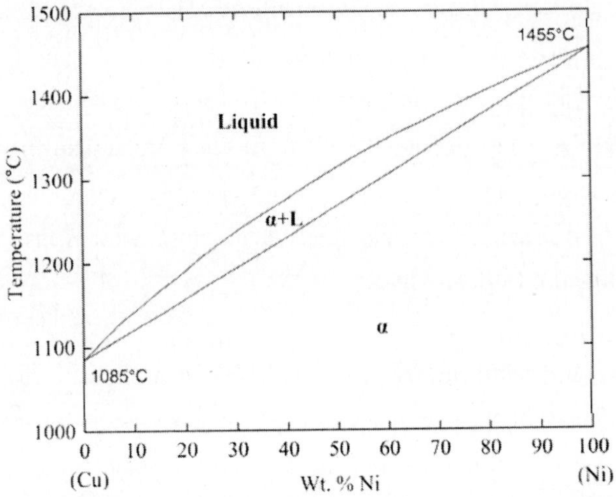

Figure 6.10 Binary phase diagram of Cu-Ni alloy

6.8.3.7 Cu-Si Alloy

Copper-silicon alloy is known as silicon bronze, which is known to have higher strength coupled with good corrosion resistance and weldability. The Silicon content in silicon bronze varies from about 1 to 4%. Increasing the Silicon content in copper-silicon alloy increases the strength, hardness, and stain-hardening behaviour while the ductility first decreases (up to 1%) and then increases with the addition of the silicon (up to 4%). Additionally, increasing silicon content decreases the electrical and thermal conductivity of silicon bronze. At high temperatures, copper silicon alloys show a hot cracking tendency.

6.8.3.8 Cu-Al alloy

Copper aluminium alloys are usually called aluminium bronze. The addition of aluminium ranges from 3 to 15% in aluminium bronzes, along with a small amount of iron, nickel, manganese, and silicon. The aluminium content directly affects the microstructure and type of aluminium bronze. Accordingly, the aluminium content determines the mechanical properties and response to the heat treatment of aluminium bronze. Adding up to 7% of aluminium forms a single alpha-type bronze that doesn't respond to heat treatment. However, these alloys show a hot cracking tendency under stressed conditions.

Further, the higher addition of aluminium in aluminium-bronze produces alpha-beta phases in the microstructure, also called duplex aluminium bronze. The presence of Alpha Beta phases in aluminium bronze makes it precipitation-hardenable. The aluminium bronze source has significant similarity with the hardenable steel, which on heating to high temperature followed by rapid cooling, results in a martensite-like microstructure. Alpha Beta phases in aluminium bronze increase the plastic range and resistance to hot cracking. Annealing of the aluminium bronzes increases the resistance to de-aluminification in salty environments.

6.8.4 Manufacturing and Copper Alloys

Casting

Casting is one of the most commonly used route for brasses to manufacture bearings, industrial valves and fittings. Shell casting, investment casting, continuous casting and centrifugal casting is the widely used to process brass. The choice of the casting method, however, depends on the number of castings, the economy of the process, finish and the dimensional tolerances desired.

Welding

The welding of copper and its alloys is considered a little difficult primarily due to the thermal and electrical conductivity. The fusion welding of the copper alloys imposes difficulty in melting the faying surfaces as the heat supplied for melting using the suitable heat source dissipated rapidly to the underlying base metal due to high thermal conductivity. Therefore, relatively high energy density welding processes (GTAW, GMAW, PAW, LBW, and EBW) are more effective in facilitating the fusion of the wing surfaces of the copper alloy base metals. A similar kind of problem is also observed during resistance welding, where-in high thermal and electrical conductivity both create difficulties in developing the weld joint. The interface heat generation due to the electrical resistance heating (joule heating) was reduced significantly due to the high electrical conductivity of the copper. Therefore, very high welding currents are needed for developing the resistance welding joints of copper and its alloy. Further, the heat generated at the joint interface by joule heating during the resistance welding also dissipated rapidly due to the high thermal conductivity, which makes developing the weld nugget difficult. Brass joined by mechanical, chemical, and metallurgical joining methods. Mechanical joining includes bolting, riveting, and crimping. Adhesives bonding is an example of chemical joining. Metallurgical joining includes welding, soldering and brazing. Brasses are readily weldable, those containing Pb, Bi, Se. Tungsten inert gas welding (TIG) and Metal inert gas welding (MIG) are generally used to weld Cu alloys. Ar, He or mixtures of the two are commonly used as shielding gases for TIG and MIG welding of Cu alloys. Non-traditional welding processes such as electron beam and laser welding produce sound joints in copper alloys. Soldering and brazing can be used to join all categories of Cu alloys. Shielded metal arc welding and resistance welding are not recommended for joining copper alloys.

6.9 Aluminium and aluminium alloys

The aluminium is the third most important metal after iron and copper due to high strength to weight ratio, good corrosion resistance, high electrical and thermal conductivity, and ease of recycling (Table 6.6). Aluminium and its alloys are relatively soft, ductile, and low strength, therefore, it is easier process by casting, rolling, extruding, welding, and machining into a variety of shapes. Aluminium is lighter (2.7g/cm^3) than the steel (7.83g/cm^3) and copper (8.93g/cm^3). Therefore, it is preferred in the automotive and transport sector. The formation of thin refractory aluminium oxide (~6.35nm) on the skin of aluminium acts protective coating against air, moisture and chemical attack. Thermal expansion of the aluminium alloys is approximately twice of steel. Therefore, contraction residual stresses associated with aluminium (about 6% by volume) frequently cause distortion and cracking. Aluminium alloys being relatively cheaper, coupled with good electrical and thermal conductivity, finds many applications such as high torque electric motors, transformer windings, heat exchangers, electrically heated appliances, and automotive cylinder heads.

Table 6.6 Physical Properties of Aluminium

Density / Specific Gravity (g.cm^{-3} at 20 °C)	2.70
Melting Point (°C)	660
Specific heat at 100 °C, cal.g^{-1}K^{-1} (Jkg^{-1}K^{-1})	0.2241 (938)
Latent heat of fusion, cal.g^{-1} (kJ.kg^{-1})	94.7 (397.0)
Electrical conductivity at 20°C (% of international annealed copper standard)	64.94
Thermal conductivity (cal.sec^{-1}cm^{-1}K^{-1})	0.5
Thermal emissivity at 100°F (%)	3.0
Reflectivity for light, tungsten filament (%)	90.0

6.9.1 Aluminium Alloys

Pure aluminium is relatively weaker (49 MPa) than alloy (700 MPa). Suitable alloying and heat treatment of alloy increase strength significantly.

There are two major categories of aluminium alloys cast alloys and wrought alloys. A further grouping of aluminium alloy based on strengthening mechanisms such as solid solution strengthening, precipitation and work hardening alloys. The addition of a few alloying element forms a solid solution with aluminium to strengthen the alloy. Aluminium alloys responding to thermal treatment called heat treatable alloys. A common heat treatment of aluminium alloys involves solutionizing, quenching, precipitation or age hardening. The work-hardening alloys gain strength from cold work, e.g. by rolling, forging, extrusion etc. Both Wrought / cast alloys can be heat-treatable or non-heat treatable. Aluminium alloy is designated using a four-digit number, with a further letter and number indicating the alloy's temper or condition.

6.9.1.1 Wrought Alloy Designation System

The first digit of the designation indicates the main alloying constituent. The second digit shows variants of the initial alloy. The third and fourth digits are related to individual alloy variations otherwise, those two have no significance but makes it unique only.

- 1xxx: Pure Aluminium (99% or more)
- 2xxx: Al-Cu Alloys
- 3xxx: Al-Mn Alloys
- 4xxx: Al-Si Alloys
- 5xxx: Al-Mg Alloys
- 6xxx: Al-Mg-Si Alloys
- 7xxx: Al-Zn Alloys
- 8xxx: Al-other elements
- 9xxx: Unused series

6.9.1.2 Cast Alloy Designation System

Similar to wrought alloys, the first digit indicates the main alloying element. The second & third digits are for specific alloy designation only. Forth digit: 0 for as-cast and 1, 2 for ingot designation.

- 1xxx: Pure Aluminium (99% or more)
- 2xxx: Al-Cu Alloys
- 3xxx: Al-Si-Cu/Mg Alloys

- 4xxx: Al-Si Alloys
- 5xxx: Al-Mg Alloys
- 6xxx: Unused series
- 7xxx: Al-Zn Alloys
- 8xxx: Al-Sn
- 9xxx: Al-Other elements

6.9.2 Effect of Alloying Elements on Properties of Aluminium

(A) **Silicon**

Al-Si alloy forms a eutectic system (12.6%Si), as shown in Fig. 6.11. Addition of silicon in aluminium improves strength to weight ratio and corrosion resistance, and wear resistance while thermal expansion co-efficient decreases (Fig. 6.12). Silicon can appear in two forms a) eutectic silicon as fine and fibrous needles 2) primary silicon as large cuboids particles as per silicon content and cooling imposed during solidification. Fine eutectic and primary silicon particles increase the mechanical properties of Al-Si alloys. Large silicon particles reduce tensile and fatigue strength due to high-stress concentration at the particle-matrix interface.

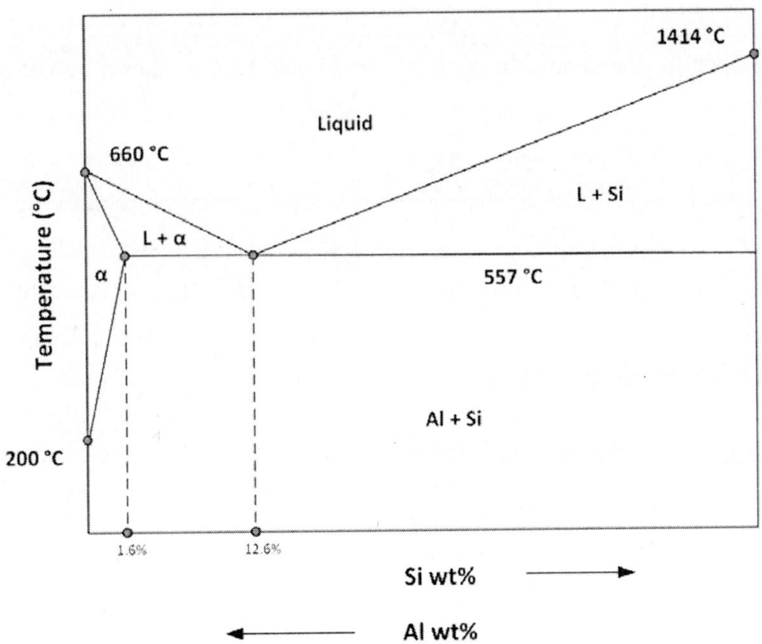

Figure 6.11: Binary phase diagram of Al-Si alloy

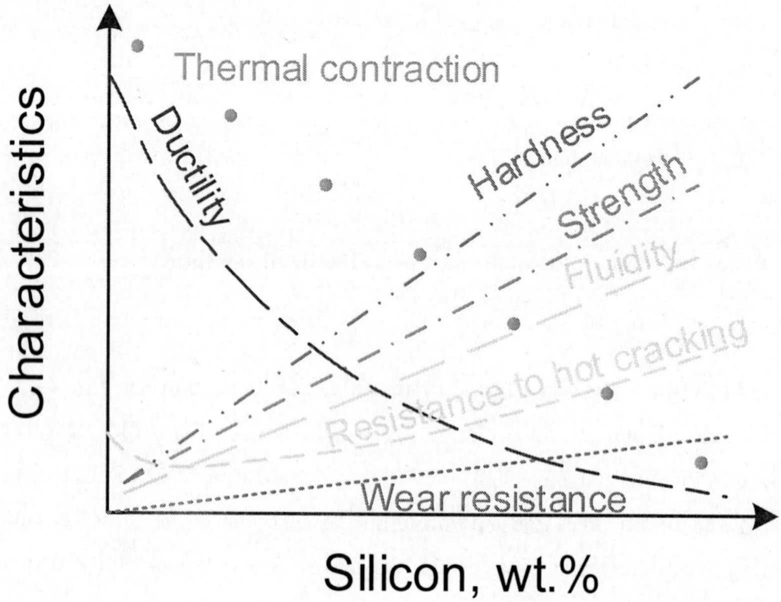

Figure 6.12: Effect of silicon on mechanical and tribological characteristics of Al-Si alloy

(B) **Copper**

Al-Cu alloy also forms a eutectic system (32.0% Cu), as shown in Fig. 6.13. The addition of copper in aluminium increases the hardness, strength, machinability, corrosion & wear resistance while fluidity, high temperature strength and ductility are compromised (Fig. 6.14). Copper can form a solid solution or inter-metallic compounds such as $CuAl_2$ (provided Mg is absent), $Al_8Cu_2Mg_8Si_6$. The addition of 4% copper decreases fluidity and wider solidification temperature range, increasing hot cracking tendency. Copper above the solubility limit forms $CuAl_2$, which improves hardness and strength but at the cost of ductility. Copper acts as a refiner for aluminium. However, copper addition reduces general corrosion resistance.

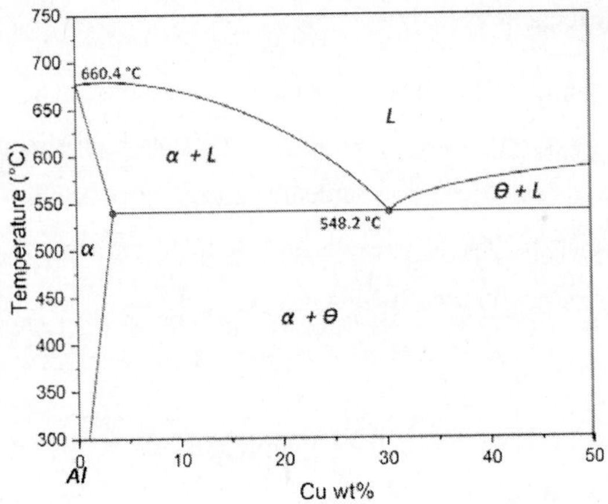

Figure 6.13: Binary phase diagram of Al-Cu alloy

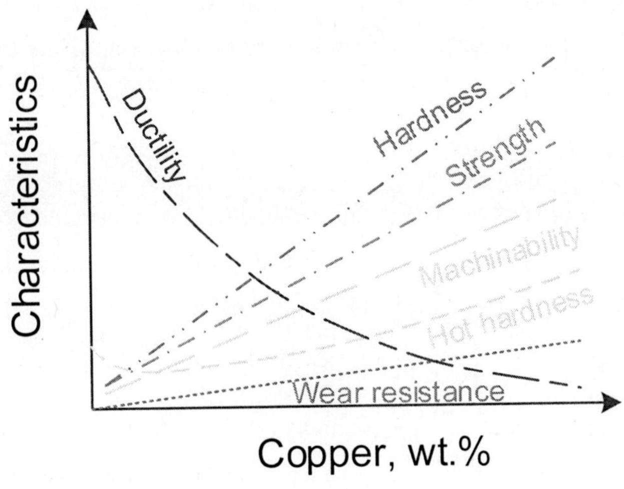

Figure 6.14: Effect of copper on mechanical and tribological characteristics of Al-Cu alloy

(C) **Magnesium**

The addition of Mg (2-12%) causes hot tearing at low Mg levels (<4%) due to a wider solidification temperature range. Magnesium (in aluminium alloy) reacts with oxygen and forms finely dispersed inclusions, reducing molten metal's fluidity and mechanical properties of the casting. The addition of magnesium (0.2 wt. %) in aluminium improves yield strength, hardness and stiffness. The higher magnesium content (>0.3%) lowers the ductility, impact energy and fracture toughness.

Mg forms eutectic (Mg_5Al_8). Magnesium increases the response of heat treatment of aluminium (Al-Cu, Al-Si, Al-Zn, Al-Mn) alloy and helps to develop high strength Al-Si alloys with an element like copper, nickel. Mg_2Si is a hardening phase with a solubility limit of approximately 0.70% Mg. Less than 0.20% Mg results in a limited amount of precipitation hardening magnesium-silicide (Mg_2Si) in Al-Si-Mg alloys (Fig. 4.4).

(D) Nickel

The addition of nickel (< 3%) in aluminium (in the presence of copper) increases the tensile properties and hardness of aluminium at elevated temperatures besides reducing the coefficient of thermal expansion coefficient. The formation of Ni-Al compound $NiAl_3$, increases resistance to thermal softening.

(E) **Phosphorus**

Phosphorous (~0.015%) is used as a refiner for primary silicon in hypereutectic Al-Si alloys. Phosphorous form aluminium phosphide (AlP), which acts as nucleants for primary silicon and produce fine primary silicon particles in the eutectic matrix

(F) Strontium

Strontium (0.008 to 0.04%) acts as a modifier for refining the morphology of eutectic in Al-Si alloys to produce a fine fibrous eutectic structure. Over-modification due to higher Sr addition causes porosity in sand mould casting processes.

(G) Titanium

Titanium (0.15%) is used as a refiner to produce fine aluminium grains and is commonly used with other elements (B, TiC). Titanium combines with boron to form TiB_2 for effective grain refinement. Excessive titanium (0.15% to 0.25%) helps to reduce hot cracking, but higher than 0.25% makes the machining difficult.

(H) Boron

Boron reacts with Ti to form titanium boride (TiB_2), which provides nucleation sites to form grain-refining phases such as $TiAl_3$. However, machinability of aluminium alloys is reduced in presence of hard metallic boride particles. Excessive boron causes furnace slugging, particle agglomeration, and increased risk of inclusions.

(I) Iron

Aluminium alloys frequently contain iron as alloying element/impurities. Iron reduces ductility and fracture toughness (Fig. 4.8a, b) while reducing the die soldering tendency. The addition of iron also appreciably decreases the thermal and electrical conductivity depending on the morphology of the IMC (Al-Si-Fe) formed. These compounds may appear as Chinese script (like $FeSiAl_5$), needle/ platelets ($FeSi_2Al_4$), angular globules, or petal like particles. Rapid cooling generally causes refinement. The effect of iron can be reduced using

1) Rapid cooling retains iron in the solid solution and improves distribution.
2) Superheating the melt changes morphology of needle shape IMC to the Chinese scripts.
3) Minor addition of alloying element like Mn, Cr, Be, Co, Mo, Ni, S, V, W, Cu, Ca etc.

(J) **Zinc**

Zn in the presence of Mg may makes aluminium alloy precipitation hardenable by forming $MgZn_2$, however, reduces the high temperature strength and hot tearing resistance.

(K) **Chromium**

Chromium acts iron corrector. Cr increases strength corrosion resistance, quench-sensitivity and reduces ductility. Chromium with aluminium forms $CrAl_7$. This compound restricts grain growth during aging. Chromium causes poisoning for grain refiners so, degrading their effectiveness.

(P) Manganese

The addition of Mn in cast Al-Si alloys improves the strength, high temperature properties and fatigue resistance. Mn is added (not more than half of the Fe %) as an iron corrector. Mn neutralizes effect of iron by forming Al_{15} (Fe, Mn)$_3$ Si_2 compound as Chinese script and minimizes the formation of needle shaped $FeSiAl_5$.

6.9.3 Non-heat treatable aluminium alloys

The non-heat treatable aluminium alloys are strengthened by solid solution strengthening, dispersion hardening, strain hardening, and grain refinement. The optimal addition of alloying elements such as silicon, iron, manganese and magnesium, increases the strength. The most effective alloying element causing solution strengthening of aluminium is magnesium. Therefore,

Al-Mg-Mn alloys (5XXX series) offer high strength even in the annealed condition. The non-heat treatable aluminium alloys are generally work hardenable (to a limited extent). Exposure to heat during manufacturing, however, reduces the work hardening effect and improves ductility.

6.9.4 Heat treatable aluminium alloys

The heat treatable aluminium alloys (2XXX, 6XXX and 7XXX series) gain strength from solid solution strengthening, work hardening and precipitation strengthening. These alloys (in annealed condition) offer almost matching strength with non-heat treatable alloys due to presence of alloying elements such as copper, magnesium, zinc and silicon. However, heat treatable aluminium alloys are primarily strengthened from precipitation hardening. The precipitation hardening involves solutionizing followed by quenching and aging either at room temperature (natural aging) or elevated temperature (artificial aging) as shown in Fig. 6.15 and 6.16. Principles related to PH hardening alloys can be seen in details chapter 1, section 1.16.4. Al-Cu (2XXX series), Al-Mg-Si (6XXX series) and Al-Zn-Mg (7XXX series) are the most common type of precipitation hardenable (PH) aluminium alloys. Strengthening of PH Al alloy mainly occurs due to the formation of fine well dispersion hardening precipitates like Al_2Cu, Mg_2Si and Zn_2Mg in 2xxx, 6xxx and 7xxx series Al alloys. An optimum T6 / T4 precipitation hardening treatment, called the peak hardening condition, leads to maximum strength and hardness. Lower strength and hardness is observed in both under-aged and over-aged condition compared to the optimum condition (Fig. 6.16). The precipitation hardening sequence observed in three common PH Al alloy during the aging process is given below.

- Al-Cu : SS ~ GP ~ S' (Al_2CuMg) ~ S (Al_2CuMg)
- Al-Mg-Si: SS ~ GP ~ ß ' (Mg_2Si) ~ ß (Mg_2Si)
- Al-Zn-Mg: SS~ GP ~ γ' (Zn_2Mg) ~ γ (Zn_2Mg)

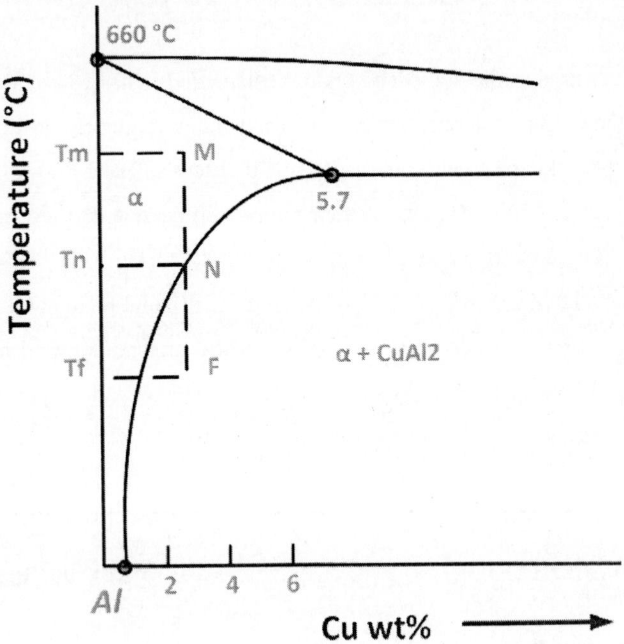

Figure 6.15: Schematic diagram showing solvus-line for Al-Cu alloy using which solution temperature "M" obtained

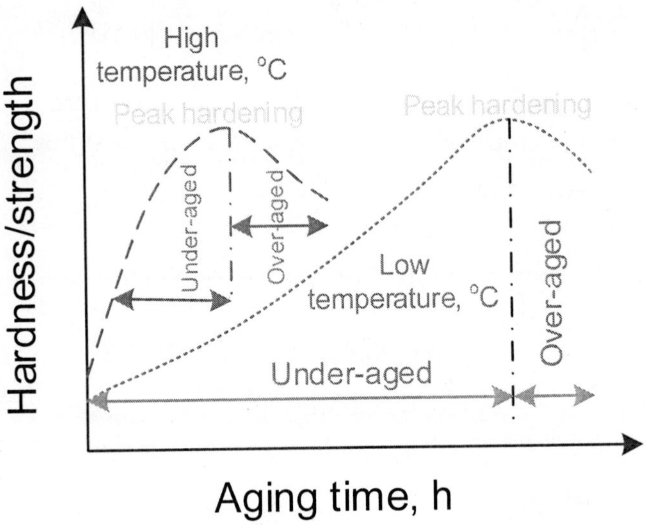

Figure 6.16: Schematic diagram showing precipitation hardening behaviour of Al-Cu alloy

6.9.5 Al-Cu-Mg alloys

The alloyed copper used in the 2000 series may be hardened to have strengths comparable to steel. These are the most popular aerospace alloys called duralumin. However, these are prone to stress corrosion cracking, so getting replaced by 7000 series PH Al-Zn-Mg alloys. Alloys in the 2xx Series - (190 to 420 MPa). Due to their high specific strength, these are popular in the aerospace sector. These contain copper from 0.7 to 6.8%. During weld fabrication by fusion welding, these show HAZ softening due to over-aging and dissolution of hardening precipitates and solidification cracking tendency due to wide solidification temperature range of few alloys (Fig. 6.17). These are welded using 4xxx series silicon or silicon and copper filler.

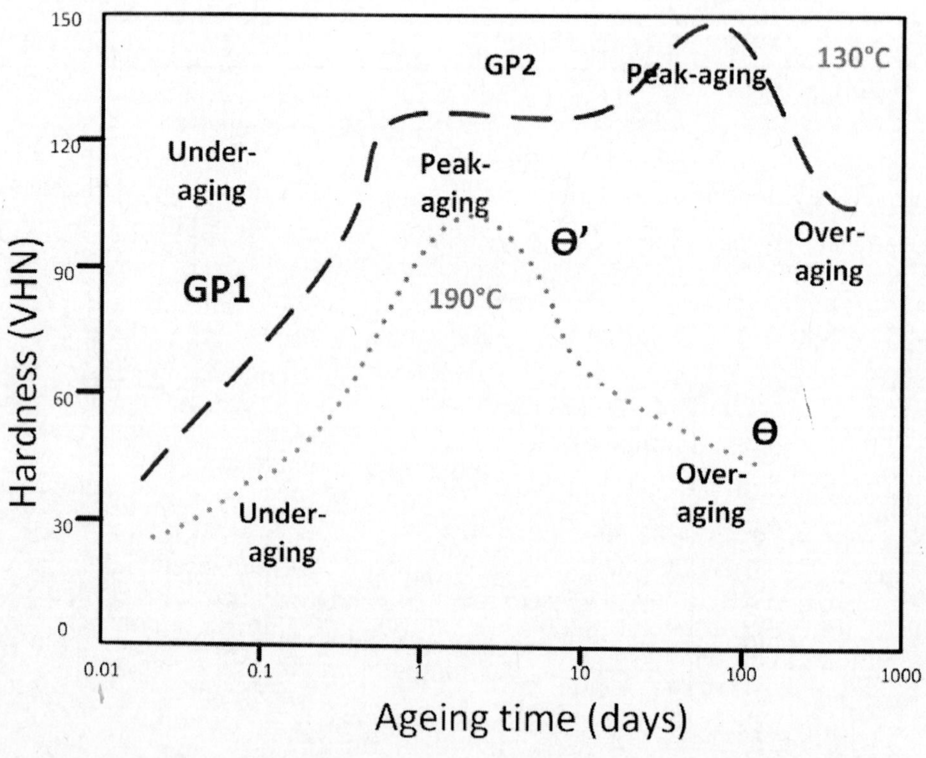

Figure 6.17: Precipitation hardening behaviour of Al-Cu-Mg alloy during artificial aging

6.9.6 Designation and Applications

The 2xxx family of aluminium-copper alloys are utilised in constructing aeroplanes, propellers, car bodywork, and screw fittings (Fig. 6.18 and Table 6.7).

Table: 6.7 Common aluminium alloys and their compositions

Designation	Si,%	Cu,%	Mn,%	Mg,%	Ni,%	Ti,%	Others,%
2011	0.4 max	5.0-6.0	-	-	-	-	Pb=0.4, Bi=0.4
2014	0.5-1.2	3.9-5.0	0.4-1.2	0.2-0.8	-	0.15 max	-
2017	0.2-0.8	3.5-4.5	0.4-1.0	0.4-0.8	-	0.15 max	-
2018	0.9 max	3.5-4.5	-	0.4-0.9	1.7-2.3	-	-
2024	0. 5 max	3.8-4.9	0.3-0.9	1.2-1.8	-	0.15 max	-
2025	0.5-1.2	3.9-5.0	0.4-1.2	-	-	0.15 max	-
2036	0. 5 max	2.2-3.0	0.1-0.4	0.3-0.6	-	0.15 max	-
2117	0. 8 max	2.2-3.0	0.2-0.5	-	-	-	-
2124	0. 2 max	3.8-4.9	0.3-0.9	1.2-1.8	-	0.15 max	-
2218	0. 9 max	3.5-4.5	-	1.2-1.8	1.7-2.3	-	-
2219	0. 2 max	5.6-6.8	0.2-0.4	-	-	0.02-0.1	V=0.1, Zr=0.18
2319	0. 2 max	5.6-6.8	0.2-0.4	-	-	0.1-0.2	V=0.1, Zr=0.18

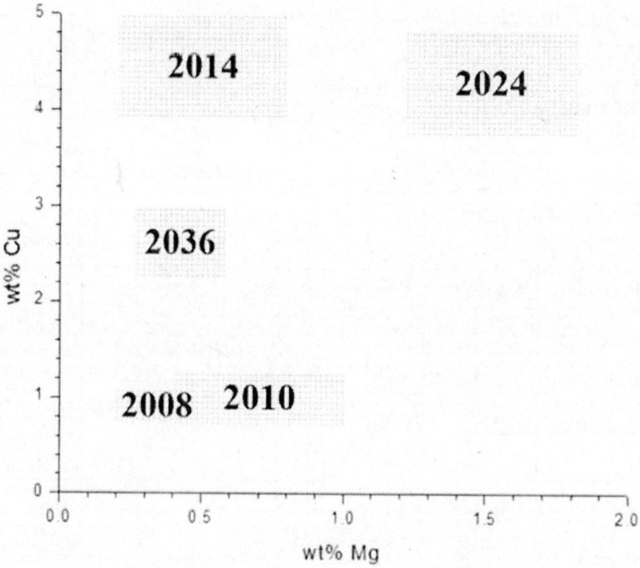

Figure 6.18: Common precipitation hardening Al-Cu alloys with designation (with Cu & Mg %)

6.10 Ni alloys

Nickel is a metal of transition group (VIII) with FCC crystal structure. Nickel alloys contain more than ten alloying elements. Nickel is frequently alloyed with cobalt, iron, chromium, ruthenium, molybdenum, rhenium, and tungsten. Alloying of nickel produces a typical microstructure in such alloys in the form of γ phase. Further, alloying of nickel with aluminium, titanium, niobium, and tantalum forms a matrix of Ni_3 (Al, Ta, Ti), and from this matrix, intermetallic compounds like Ni_3Al, Ni_3Ta, Ni_3Ti precipitate. Thus, the microstructure contains the γ phase and γ'Phase. The crystal structure of the γ phase is FCC, and the γ'phase is simple cubic only, as shown in Fig. 6.19. Alloying of nickel-iron super alloys with an element like niobium forms γ'' (BCT structure with c/a ratio 2). Coherent phases are formed when the atomic radius of alloying element is similar to the parent metal, causing significant precipitation hardening. Then the lattice parameter is found same for both the γ phase and γ'phase, resulting in a coherent structure. The coherent structures in nickel alloys increase thermal stability (in strength and microstructure) at elevated temperatures because the cross-slip is needed for the dislocation movement. Therefore, nickel alloys retain higher strength and resistance to creep at high temperatures (up to 800°C) than ferritic stainless steel (500-550$°C$), and titanium alloys (up to 700°C). To realize further higher creep resistance components of super alloys are produced in the form of single crystals as polycrystalline metals show lower creep resistance due to grain boundary (sliding) weakening at elements.

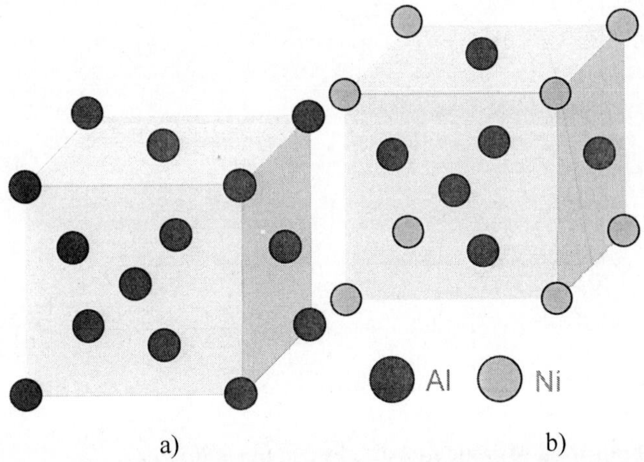

Figure 6.19: Schematic diagram showing a) Crystal structure of γ and b) Crystal structure of γ'

6.10.1 Microstructure and mechanical properties

The microstructure of nickel alloys, commonly composed of γ phase (solid solution of nickel with Co, Fe, Cr, Mo, W) and γ' phase (Ni_3Al, Ni_3Ta, Ni_3Ti) while nickel-iron alloy with Nb form γ and γ'' phase (Ni_3Nb) which are coherent with the matrix. However, unfavourable phases like μ, σ, laves, etc., are also formed due to excessive alloying with Re, W, Ta, Mo, and Ru. Nickel alloys are designed based on strengthening mechanisms like solid solution strengthening, oxide-dispersion strengthening, carbide and borides formation at grain boundaries, and precipitation strengthening.

Carbides and borides segregated at the grain boundaries enhance the creep resistance and high temperature strength as deformation is caused by climbing up mechanism for dislocation movement in the presence of hard and stable particles at the grain boundary. The microstructure (γ,γ',γ'', μ, σ, laves, carbides, borides etc.) and the corresponding composition of nickel alloys is shown in Table 6.8. Alloying elements like Cr, Al, and Ti enhance the oxidation resistance of nickel alloys. Creep resistance is best when the microstructure comprises 70% of γ'pPhase, 30% of γ phase, and it is realised through controlled alloying of nickel with Al, Ti, and Ta. The choice of grain size is complicated due to the opposite effect on creep and fatigue strength of nickel alloy (like in the case of gas turbine blades), as evident from Fig. 6.20.

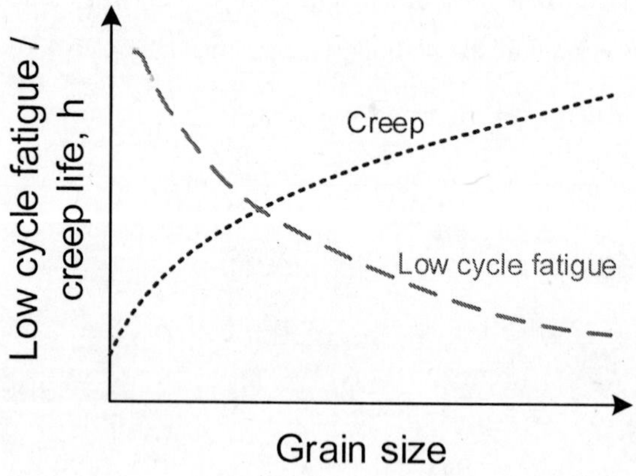

Figure 6.20: Variation of low cycle fatigue life and creep rupture life along with the grain size

Table 6.8: Chemical composition of some nickel-based super alloys (wt. %).

Common name	Ni	Cr	Co	Fe	Mo	Al	Ti	Si	Mn	C	W	Nb	Ta	B	Zr
Alloy X	47.3	22.0	1.5	18.5	9.0			0.5	0.5	0.1	0.6				
Alloy 617	54.0	22.0	12.5		9.0	1.0				0.07					
Alloy 706	41.5	16.0		40.0		0.2	1.8	0.2	0.2	0.03		2.9			
Alloy 718	52.4	19.0		18.5	3.0	0.5	0.9	0.2	0.2	0.04		5.1			
Alloy 783	28.5	3.0	34.0	26.0		5.4	0.1					3.0			
Alloy 901	42.5	12.5		36	5.8		2.9								

6.10.2 Manufacturing and Nickel alloys

Common manufacturing methods applied for nickel alloys are forging, casting, powder metallurgy, welding, machining and 3D printing. Investment casting produces a near-net shape, maintains desired composition, coarse-grained directional solidification, and single-crystal are obtained through directional solidification.

Initially, only deformation-based manufacturing processes were applied to nickel-based alloys, however, those were limited to deformable alloys only, which were unsuitable for many nickel alloys. Powder metallurgy is extensively used where forging techniques were not applicable. Fusion welding using matching filler metal is preferred when nickel alloy is largely free from hard phases. The solid-state joining technique, like diffusion bonding with suitable interlayer has been extensively used for developing dissimilar metal joints of nickel alloys with titanium and steel. In brazing, due to the low heat input feature is also used joining nickel alloy as per the suitability of joint application. The joining of precipitation-hardening nickel alloys using heat input methods is preferred due to high hot cracking and post-weld heat treatment cracking tendency.

The most common welding issues observed in fusion weld joints of nickel alloys includes low strength of HAZ, reheat cracking, and hot cracking in the partially melted zone. The low strength of HAZ can be restored by post-weld heat treatment (artificial aging). Reheat cracking occurs during the post-weld heat treatment of the joint. The precipitation of hard and brittle particles at grain boundaries before relieving residual stress causes cracking. Reheat cracking can be avoided by a) isothermal treatment, b) low heat input welding, and c) minimising impurities. Hot cracking in nickel-based alloys is mainly caused by MC carbide lave phases formed during weld metal solidification and encouraged by a wider solidification temperature range (mushy zone). Further, the coarse grain structure of nickel alloy and high restraint during welding promote hot cracking. Hot cracking can be reduced by choosing the proper filler metal to minimise the mushy zone temperature range, refining grain structure, and reducing restraint.

6.10.3 Applications

Nickel alloys are mainly used for making high-temperature components used in gas turbine power plants (turbine blades), aircraft, chemical plant equipment, and steam turbine power plant components besides automotive components, metal processing such as hot work tools, dies, casting dies, space vehicle components, medical components etc.

6.11 Titanium alloys

Titanium alloys are known for two unique properties: high specific strength (both at room temperature and high temperature) and corrosion resistance, making them attractive metals for chemical, medical, and aerospace industries. The development of titanium aluminides has resulted in improved oxidation resistance as well. Ti-Al alloys are candidate metals for choice amongst nickel alloys and Cr-Mo steels for high-temperature applications around 500°C (Fig. 6.21).

6.11.1 Physical metallurgy

Titanium is an allotropic metal having a hexagonal close-packed (HCP) structure (β-Ti) at low temperatures and a body-centered cubic (BCC) structure (α-Ti) at high temperatures. For pure titanium α to β transition temperature is 882 °C (Fig. 6.22). The allotropic behaviour of Ti as a function of temperature (like ferrous metals) helps to change properties as per need. Crystal structure affects both yield strength and diffusion rate. The hexagonal crystal lattice of α titanium causes anisotropy in mechanical properties (on basal plane and vertical plane).

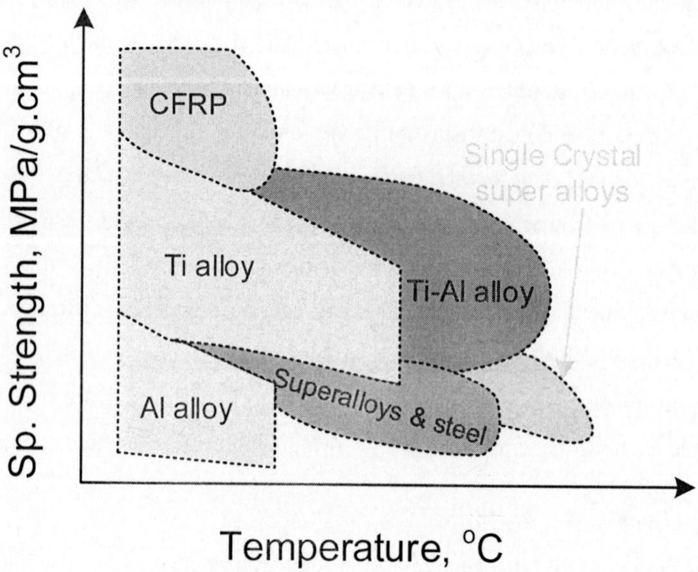

Figure 6.21: Specific strength of common metals as a function of temperature

The alloying element in titanium alloys are grouped into three categories, namely neutral, α-stabilizers, and β-stabilizers according to their effect on β to α transition temperature (Fig. 6.22). The α-stabilizing elements (Al, C, N, O) increase α to β transition temperature, while the β-stabilizing elements reduce the same, while neutral elements hardly affect α to β transition temperature (Fig. 6.22)

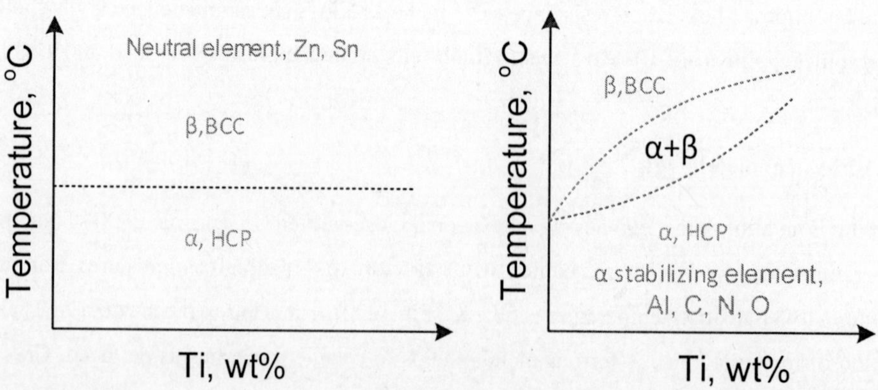

a)

b)

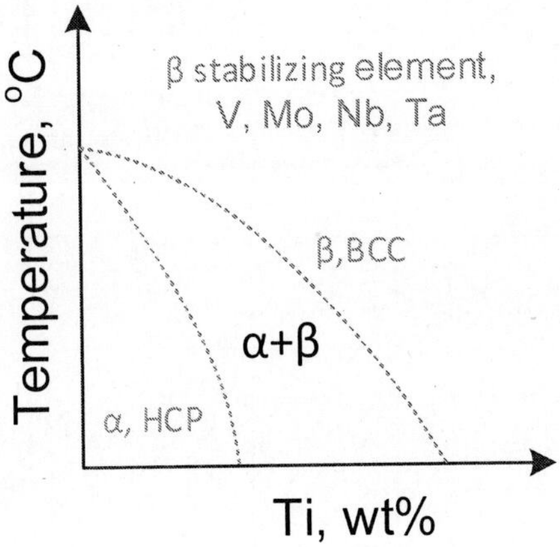

c)

Figure 6.22: Effect of different types of alloying element on *α, β* phases in Ti alloy (a) neutral, (b) α-stabilizing, and (c) β-stabilizing elements.

Aluminium is the most important α-stabilizing alloying element for titanium besides interstitial elements like carbon, nitrogen, and oxygen. The α-stabilizers result two-phase (α+β) field apart from increasing α to β transition temperature.

Based on primary phase(s), titanium alloys can be categorised as near-α, α, α+β, metastable β, and β alloys. The surface plot (3D diagram) shows the effect of α, β stabilisers and temperature on phases formed by Ti (Fig. 6.23). The α alloy can be a pure (CP) titanium or the titanium alloyed neutral and/or α-stabilizing elements while near α-alloys contain little amounts of β-stabilizing elements. The α+β alloys are the most popular Ti alloy with their volume fraction between 5 and 40% at room temperature. An increase of β-stabilizing elements (in α+β alloy) results in metastable β alloys. In metastable β alloys, β-stabilizing element transforms the β phase of two phase system i.e. α+β, into a martensite like structure on rapid cooling. These alloys still consist of equilibrium α phases (>50%) with martensite like structure. The single-phase β alloys are traditional titanium alloy of the high % of alloying of β-stabilizing elements.

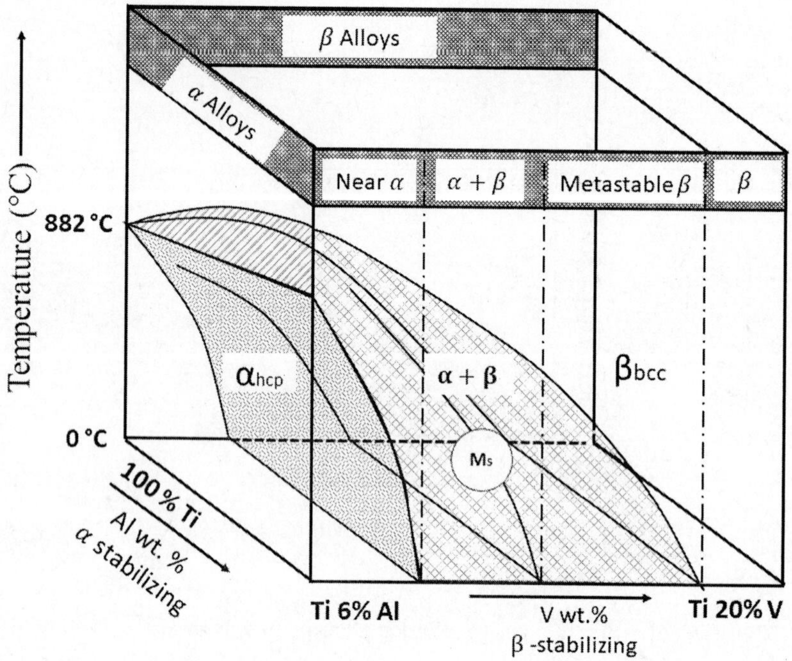

Figure 6.23: Schematic of a 3D phase diagram of Ti-Al-V alloy showing variation in phases as a function of composition and temperature

6.11.2 Mechanical properties

The most common titanium alloys are used in solution treated, artificially aged and annealed conditions. An increase of temperature significantly affects the thermal conductivity, hardness, strength and young's modulus of a common Ti-6Al-4V alloy. For example, heating of Ti-6Al-4V alloy from room temperature to 1200°C increases thermal conductivity 3 fold while hardness, strength and young's modulus decreases rapidly with the temperature increase [4].

6.11.3 Applications of Ti-Alloys:

Ti-alloys use is restricted to military applications, aeroplanes, spacecraft, bicycles, medical gadgets, jewellery, highly stressed components like connecting rods on luxury sports vehicles, prestige sporting goods, and consumer electronics due to the exorbitant cost of both raw materials and manufacturing methods. The Ti-6Al-4V alloy is utilised in biomedical components, surgical equipment, dental implants, prosthetic femoral components, and other applications in the interest of its biocompatibility and excellent corrosion and mechanical properties. Table 6.9 shows few important Ti-alloys and their applications in different sectors.

Table 6.9: Applications of Ti alloys

Ti Alloy	Applications sector		
	Aerospace Applications	Automotive Applications	Medical Applications
	Airframe structures	Frame structures	Joints
Ti-6Al-4V	Windows frames,	Armour, Body, Suspension springs	Hip and knee joints, Dental materials
CP-Ti and Ti-11Mo-6Zr-4Sn	Floors structure	Body structures	Valves and connectors of Heart, Dental materials
Ti-15V-3Cr-3Sn-3Al	Springs		
Ti-3Al-2.5V	Hydraulic tubing		
Ti-6.8Mo-4.5Fe-1.5Al, Ti-6Al-4V		Structures of suspension springs	
	Gas Turbine Engines	Automotive Engines	
Ti-6Al-4V, Ti-6-2-4-2S	Fan Discs and Blades, Compressor Discs, Compressor Blades		
Ti-35V-15Cr	Compressor Stators		
TIMETAL 21S	Nozzle Assembly		
Ti-6Al-4V		Outlet Valves, Intake Valves, Connecting Rods	
Gamma(Ti-Al), Grade-2		Turbocharger rotors, Outlet valves, Exhaust Systems	

UNIT SUMMARY

This unit presents chemical composition, microstructure, and mechanical properties of following common metals: ferrous metals (wrought iron, carbon steel, cast iron, alloy steel, stainless steel, maraging steel), copper and copper alloys (oxygen-free and oxygen-bearing copper, brass, bronze, copper-nickel, copper-silicon, copper-aluminium, copper-zinc), aluminium, aluminium alloys (cast, work hardening, precipitation hardening aluminium alloys such as Al-Cu, Al-Si, Al-Cu-Mg alloys), super-alloys, alpha-titanium, beta-titanium and alpha-beta titanium alloy. Phase diagrams of selected alloys (Al-Cu, Cu-Zn, Al-Si, Ti-Al-V) have also been presented for the sake of clarity and better understanding. Common commercial applications of these metals have also been described.

EXERCISE

Questions for self-assessment

1. Ferrous metal contains the minimum carbon content is
   a. Wrought iron
   b. Steel
   c. Cast iron
   d. All of these
2. Steel in which only the carbon content (as an alloying element) is controlled
   a. Low alloy Steel
   b. Stainless steel
   c. Plain Carbon Steel
   d. All of these
3. An element that makes stainless steel corrosion-resistant is
   a. Carbon
   b. Chromium
   c. Manganese
   d. Nickel
4. The chromium carbide precipitation in austenitic stainless steel causes
   a. Weld decay
   b. General corrosion
   c. Softening
   d. Hardening
5. Resistance to weld decay can be achieved through
   a. Adding Nb, Ti
   b. Reducing carbon content
   c. Solution treatment above 1050 degrees centigrade
   d. All of these
6. Copper is found difficult to weld by spot welding due to
   a. High density
   b. High electrical conductivity
   c. High strength
   d. All of these

7. Copper-nickel alloy showing a solidification temperature range greater than 150-degree centigrade will be sensitive for
    a. Softening
    b. Deformation
    c. Hot tearing
    d. All of these
8. A precipitation-hardening aluminium alloy subjected to over-aging during the T6 heat treatment. The alloy will show
    a. Refinement of precipitates
    b. Coarsening of precipitates
    c. Uniform distribution of precipitates
    d. All of these
9. A type of cast iron that will show maximum ductility is
    a. Grey cast iron
    b. Malleable cast iron
    c. White cast iron
    d. Nodular cast iron
10. Cast iron, usually composed of alloy carbides, iron carbide, and martensite/pearlite, is
    a. Grey cast iron
    b. Malleable cast iron
    c. White cast iron
    d. Nodular cast iron
11. The crystal structure of alpha-titanium alloy is
    a. HCP
    b. BCC
    c. FCC
    d. BCT
12. The addition of an element in titanium that increases the alpha-to-beta phase transition temperature is
    a. Zn
    b. Mo
    c. V
    d. Al

13. Element in titanium alloy, which does not affect the alpha-to-beta phase transition temperature, is
    a. Zn
    b. Mo
    c. Al
    d. V
14. The metal which offers the highest thermal stability and resistance to softening is
    a. Chromium-molybdenum Steel
    b. Titanium alloys
    c. Polycrystalline Nickel alloys
    d. Single crystal super alloys
15. In general, an increase in grain size of metals increases
    a. Tensile strength
    b. Fatigue strength
    c. Creep strength
    d. Yield strength

Answers of Multiple Choice Questions

Key for MCQ: 1 a, 2 c, 3 b, 4 a, 5 d, 6 b, 7 c, 8 b, 9 d, 10 c, 11 a, 12 d, 13 a, 14 d, 15 c

Short and Long Answer Type Questions

1. Why do we need different metals for making components for different service conditions?
2. Enlist the residual elements present in plain carbon steel as traces.
3. How does the sulphur affect the ease of casting and welding of steel?
4. What is the effect of elevated inclusions present in wrought iron on the mechanical properties and cracking tendency?
5. Why does wrought iron offer high percentage of elevation during the tensile test?
6. Based on the carbon content, give the classification of the carbon steels.
7. What makes stainless steel corrosion-resistant?
8. Write about common types of stainless steel and general mechanical and metallurgical characteristics.
9. How does the following element affect the microstructure and mechanical properties of steel?
    a. Tungsten (W)

b. Chromium (Cr)
c. Silicon (Si)
d. Molybdenum (Mo)
e. Vanadium (V)
f. Nickel (Ni)
g. Manganese (Mn)
h. Cobalt (Co)
i. Titanium (Ti)
j. Aluminium (Al)

10. How does alloy steel offer different mechanical properties and is of hardening than carbon steels?
11. What is the significance of carbon, chromium, and nickel equivalent steels?
12. What are the common problems encountered during the fusion welding of austenitic stainless steel?
13. Explain the approached to minimize the problem of weld decay.
14. Describe the common applications of the different types of steel like carbon steel, alloy steel, and stainless steel.
15. What are the typical properties needed for making the tools and dies?
16. Distinguish the water-hardening tool steels and shock-resisting steel.
17. Describe the strengthening mechanisms active in managing steel leading to high strength.
18. What do you understand from the embrittlement of the maraging steel?
19. Enlist the common types of cast irons.
20. Describe the microstructure, and mechanical properties of the following types of cast irons:
    a. Grey cast iron
    b. Malleable cast iron
    c. Nodular cast iron
    d. White cast iron
21. Explain the role of silicon in cast iron in the microstructure and mechanical properties.
22. How does the form of carbon and morphology of graphite affect the mechanical properties of cast irons?
23. How does the electrical conductivity of metals expressed with respect to copper?

24. What is brass?
25. How does zinc content affect the appearance, mechanical properties, and ease of manufacturing of brass by casting and welding?
26. What is the effect of alloying elements? on solidification cracking tendency of copper alloys?
27. Why does the oxygen-bearing copper show a cracking tendency in a hydrogen environment?
28. What is the effect of the Alpha/Beta phase in copper alloys on the hot working characteristics?
29. What are the properties of copper affect=ting the ease of fabrication by welding?
30. How does the following element affect the microstructure and Mechanical properties of aluminium alloys?
    a. Boron (B)
    b. Copper (Cu)
    c. Silicon (Si)
    d. Strontium (Sr)
    e. Vanadium (V)
    f. Zinc (Zn)
    g. Iron (Fe)
    h. Magnesium (Mg)
    i. Titanium (Ti)
    j. Manganese (Mn)
31. Compare the under-ageing, over-ageing and peak hardening related to the precipitation hardening alloys.
32. How does the grain size affect the low cycle fatigue resistance and the creep resistance of super alloys?
33. Write the significance of neutral elements, alpha and beta stabilising elements in Titanium alloys.
34. How do the neutral, alpha and beta-stabilizing elements affect the stability of the Alpha phase?
35. Explain the phase transformations in Ti-Al-V alloys as a function of composition and temperature using suitable schematic diagram.

PRACTICAL

1. Conduct hardness/toughness test on common ferrous metals (carbon steel, alloy steel and cast iron) aluminium and copper alloy and comment
2. Conduct a corrosion test by dipping austenitic stainless steel and carbon steel samples (of the same size) in saline water for 24 hours. Observe staining on the steel samples and measure the weight gain due to corrosion.

KNOW MORE

Observe the many engineering components and systems in the public domain like electric poles, steel bridges, railway tracks, etc., and comment on the general condition about appearance, and corrosion condition. Try to learn about the metals used for making the components, systems, and parts used by human beings in our daily life. Explore the various field components used by us and investigate the possible causes of failure and their general condition.

REFERENCES AND SUGGESTED READINGS

1. Callister, Materials science and engineering, Wiley, Publication (2014)
2. S D Avner, Introduction to Physical Metallurgy, McGraw Hill, Publication (2002)
3. D K Dwivedi, Fundamentals of Metal Joining, Springer nature, Singapore (2022)
4. D K Dwivedi, Surface engineering, Springer nature, Germany (2018)
5. D K Dwivedi, Production and properties of cast Al-Si alloys, New Age International (2013)
6. D K Dwivedi, NPTEL Course, Fundamentals of Manufacturing Processes, https://archive.nptel.ac.in/courses/112/107/112107219/
7. Structure/Property Relationships in Irons and Steels, ASM International, Second Edition J.R. Davis, Editor, (1998), 153-173
8. American welding society, welding of coper and copper alloys (1997), 7-11
9. American society for metal (International), Ferrous metals, Volume 1, (1993)
10. Leyens, Christoph, and Manfred Peters, eds. Titanium and titanium alloys: fundamentals and applications. John Wiley & Sons, 2003.
11. Veiga, Celestino, J. P. Davim, and A. J. R. Loureiro. "Properties and applications of titanium alloys: a brief review." Rev. Adv. Mater. Sci 32, no. 2 (2012): 133-148.
12. Donachie, M.. "Titanium: A technical guide((Book))." Metals Park, OH, ASM International, 1988, 484 (1988).

Dynamic QR Code for the Unit 6

REFERENCES FOR FURTHER LEARNING

Each unit is provided with reference and suggested books for further reading/learning

CO AND PO ATTAINMENT TABLE (To be incorporated by Author)

Course outcomes (COs) for this course can be mapped with the programme outcomes (POs) after the completion of the course and a correlation can be made for the attainment of POs to analyse the gap. After proper analysis of the gap in the attainment of POs necessary measures can be taken to overcome the gaps.

Table for CO and PO attainment

Course Outcomes	**Expected Mapping with Programme Outcomes** *(1- Weak Correlation; 2- Medium correlation; 3- Strong Correlation)*											
	PO-1	PO-2	PO-3	PO-4	PO-5	PO-6	PO-7	PO-8	PO-9	PO-10	PO-11	PO-12
CO-1	3	2	3	2	1	-	-	-	-	-	-	-
CO-2	2	2	3	2	2	-	-	-	-	-	-	-
CO-3	2	2	3	3	2	-	-	-	-	-	-	-

INDEX